AF513497

DIFFERENTIAL CALCULUS WITH EQUATION

DIFFERENTIAL CALCULUS WITH EQUATION

H.R. Mittal

CYBER TECH PUBLICATIONS
4264/3, Ansari Road, Daryaganj, New Delhi-110002 (India)
Ph.: 011-23244078, 011-43559448 Fax: 011-23280028
E-Mail: cyberpublicationsdelhi@yahoo.com
Website: www.cybertechpublications.com

DIFFERENTIAL CALCULUS WITH EQUATION

H.R. Mittal

© Reserved

First Edition 2010

All rights reserved. No part of this publication may be reproduced, stored in a retrieval system or transmitted, in any form or by any means, electronic, mechanical, photo copying, recording or otherwise, without the prior permission of the publisher.

Published by : Asian Offset Printers

G.S. Rawat for **Cyber Tech Publications**
4264/3, Ansari Road, Daryaganj, New Delhi-110002 (India)
Ph.: 011-23244078, 011-43559448 Fax: 011-23280028
E-Mail: cyberpublicationsdelhi@yahoo.com
Website: www.cybertechpublications.com

PREFACE

The Subject matter has been discussed in such a simple way that the students will find no difficulty to understand it. The proofs of various principles and examples has been given with minute details.

Each chapter of this book contains complete theory and fairly large number of solved examples. Sufficient problems has also been selected from various universities examination papers.

The author shall be gratefully to the readers who point out errors and omissions which inspite of all care might have been there. The subject matter has been presented in a simple and systematic manner. To familiarize the student about expected type of questions in the examinations, important summary and review questions have been provided at the end of each chapter.

The author will feel amply rewarded if the book serve the purpose for which it is meant. Suggestions for further improvement of this book are most welcome and would be incorporated in the next edition.

— Author

Contents

1

Complex and Real Numbers with Functions

INTRODUCTION

Complex numbers are those numbers which contain its one part as real and another part as imaginary, while both the numbers involved in the complex number belong to R. More general complex numbers are defined by a + ib, or x + iy. & a, b, x, $\in$ R, such as i = $\sqrt{(-1)}$, so it becomes as imaginary part.

Complex numbers obeys the same algebraic rules as the ones for real numbers. Therefore, the complex form a number system. The "zero" denoted by o, is the number of oi, the number "one" denoted by 1, is the number 1+ o.i. In general we can identify the number a + o.i with a real number a; that is we can consider a complex number system that includes all real numbers

The sum, difference, product and division of complex number is again a complex number.

$$(a + ib) + (c + id) = (a + c) + i (b + d)$$

$$(a + ib) - (c + id) = (a - c) + i (b - d)$$

$$(a + ib) . (c + id) = (ac + bd) + i (ad + bc)$$

$$\frac{(a+ib)}{(c+id)} \times \frac{(c-id)}{(c-id)} = \frac{(ac+bd)+i(bc-ad)}{c^2+d^2}$$

$$\frac{(a+ib)}{(c+id)} \times \frac{(c-id)}{(c-id)} = \frac{ac+bd}{c^2+d^2} + i\frac{bc-ad}{c^2+d^2}$$

The number x - iy, symmetric to z = x + iy in the x - axis, is called the conjugate of z = x + iy and denoted by .

Complex numbers can be respresented graphically by the points in the xy - plane, the nubmer z = x + iy Being represented by the points (x, y). The distance from origin to the points (x, y) is called the absolute value or modulus of the complex number z = x + iy and denoted by /x + iy/ or /z.

Thus /z/ = $\sqrt{x^2 + y^2}$, i.e. r = $\sqrt{x^2 + y^2}$ and if θ is the inclination of straight line in the positive direction of x-axis then angle θ is called the argument (or amplitude) of complex number z x + iy.s

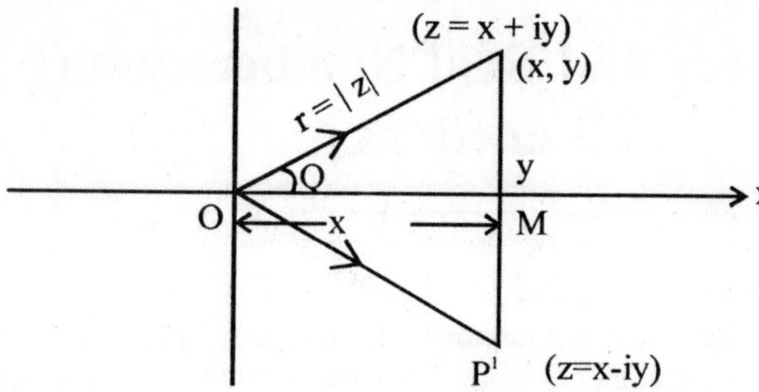

Fig. 1

From plane to trigonometry

$$x = r \cos \theta, \ y = r \sin \theta.$$

$$\therefore \tan q = \frac{y}{x} \Rightarrow \theta = \tan^{-1}\left(\frac{y}{x}\right)$$

The Complex number z = x + iy can also be represented as z = r (cos θ + i Sin θ), which is known as polar or Trigonometrical form of complex number z = x + iy.

RATIONALIZATION OF COMPLEX NUMBER

The process to change the imaginary part of any complex number to real number is known as Rationalization of complex number. Similarly, the process to change the imaginary part involved in the denominator of any complex number to real number is called Rationalization of the denominator of the complex number. This can be done only by multiplying the given complex number through its conjugate.

For example, Rationalize the denominator of the complex number Given,

$$\frac{1}{3+4i} \text{ Given } \frac{1}{3+4i} = \frac{1}{3+4i} \times \frac{3-4i}{3-4i} = \frac{3-4i}{(3)^2 - (4i)^2}$$

$$= \frac{3-4i}{9+16}$$

$$= \frac{3-4i}{25}$$

THE REAL NUMBERS

The real numbers arise naturally in measuring distances or more generally in giving the relative position of points on a line. We choose a reference point O on the line, choose a unit distance, and then indicate the position of each point on the line in terms of the chosen unit distance from O. All the points which lie on the right hand side of the reference point (origin) have plus sign and the numbers on the other hand have negative signs. The result as the familiear number axis or number line. Every real number is to be represented by one point on the axis.

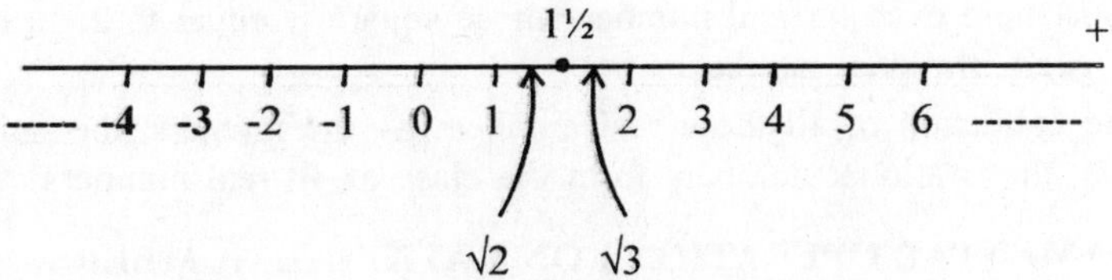

Fig. 2 : Number axis

Every pont on the axis represents one number. The point O itself represent the number O (zero), the point at one unit from O corresponds to the number +1 and -1, and the points two unit from O Corresponds to the numbers 2, -2 and so on.

The numbers that correspond to a whole number of units on the plus side (R.H.S) are the positive integers: 1, 2, 3, ... 10, ...; those that correspond to a whole number of unit on the negative (L.H.S) side are the negative integers : –1, -2, -3, -4, ... –10, ...

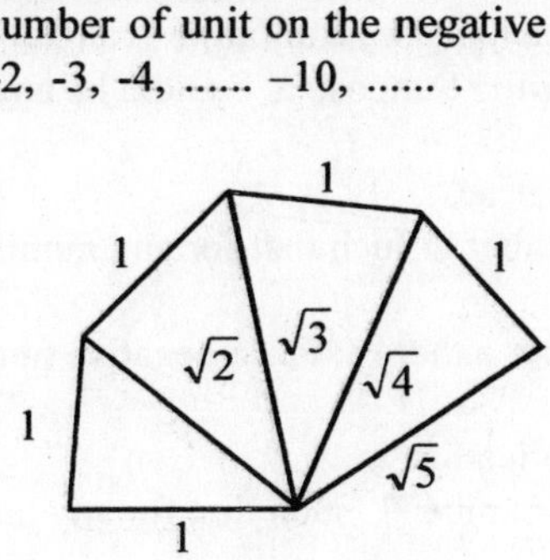

Fig. 3: Irrational numbers

The number 0, 1, -1, 2, -2, 3, -3, ... are called the integers. Again any line segment can divided into a given number of equal parts, there we have fractional numbers which can be written in the form , where m and n are integers and n is positive. These numbers are known as rational numbers. A real number that is not rational is said to be irrational.

The question arises whether corresponding to every point on the real line there is a unique rational number. We shall see that it is not so.

We now prove that there exist no rational number whose square root is equal to 2. If possible, let be a rational number whose square is equal to 2, where p, q are positive integers and that they are prime to each other. thus, or $p^2 = 2q^2$. so that p^2 is an even number and therefore p is an even number. Since square of an odd number is an odd number. Hence p = 2k and $p^2 = 4k^2$ i.e. $2q^2 = 4k^2$ or $q^2 = 2k^2$, where k is a positive integer. q^2 and consequently q is also an even and hence p and q have a common factor which contradicts our hypothesis that p and q are prime to each other.

Thus there exist no real number whose square is equal to 2. In other words, is an irrational number.

The collection of all these real numbers — the integers, the rational numbers, the irrational numbers form the class of all real numbers.

FUNDAMENTAL OPERATIONS ON RATIONAL NUMBERS

If a and b are any two rational numbers, then the basic laws of operations viz. addition (subtraction) and multiplication (division) when applied to rationals gives again rationals so that a + b, a - b, ab, are also rational numbers. Further, the operations addition and subtraction satisfy the following laws.

1. *Commutative Law :* If a and b are any rational number, then a + b = b + a and ab = ba
2. *Associative law :* For any three rational numbers a and b, c. (a + b) + c = a + (b + c) and (ab) c = a (bc)
3. *Distributive law :* Suppose a, b and c be any three rational numbers, then
   a (b + c) = ab + ac.
4. There is a number O such that for any number a,
   a + o = a.
   For any number a there exists a negative number -a, such that a + (-a) = o.

   For multiplication

6. There is a number 1. such that for any number a
   a . 1 = a.
7. For any number a ≠ o, there exists a number such that .

The number b is called the reciprocal of a. From the above rules we can show that, we can subtract any two numbers and can divide any two numbers, except for division by zero. Furthermore, ab = o, iff a = o or b = o.

Aslo (-a) (-b) = ab , (-a) b = - (ab)

Since the collection of all real numbers obeys the laws form 1 to 7, we can state that the real numbers form a number system.

FUNCTIONS

The idea of a function is essential in Economics. The term function was first used by Leibnitz to denote the dependence of one quantity on another. For example, the price of a product and the quantity demanded for the item; or the demand of a product and a seller's revenue. Suppose that to each element of a set A, there is assigned, by some manner or other, a unique element of B. We name this assignment as function or mapping. Thus, *a function is a rule that assigns to each element in a set A, one and only one element of a set B.*

If f denotes the rule, we write,

$$f : A \rightarrow B.$$

The set A is called the *domain* of the function. If x is an element in the domain of a function f, then the element that f associates with x is denoted by the symbol $f(x)$. $f(x)$ is the value of f at x and is called the *image* of x under f. The set of all possible values of $f(x)$ as x varies over the domain is called the *range* of f. The fact that y is a function of x is expressed symbolically by the equation:

$$y = f(x)$$

Other letters such as F, g, h or Greek letters such as ϕ, ψ, θ are also used to express functions. When we write $y = f(x)$, x is termed as the *independent variable* and y as the *dependent variable.*

A set can be expressed either by a set of ordered pairs, or by some algebraic formulae or by some statements. In general, the domain or co-domain of a function need not be sets of real numbers. But functions with their domain and co-domain as sets of real numbers have special importance. Functions of this type are called *real functions* or real valued functions. We shall, however, simply use the word function to mean a real function.

TYPES OF FUNCTIONS

Mainly the functions are of two types given below :

(1) **Algebraic Functions:** Those functions which are the result of performing upon the variable x and any number of constants, a

finite number of operations of addition, subtraction, multiplication, diversion and root extraction are called an algebraic functions. Algebraic functions are as follows:

(i) *Linear Function:* A function Y = mx + c, is said to be linear function, where m and c are fixed real numbers. The graph of each function is a straight line of slope m.

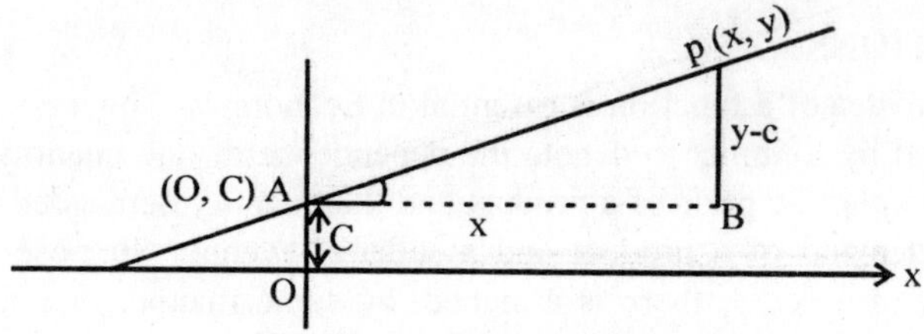

Fig. 4: Graph of linear function

C is called in intercept of the straight line with axis-y.

(ii) *Single-valued and Multiple-valued functions:* If for each value of x there corresponds a single value of y, then the function is said to be a single-valued function, however; if for each value of x there corresponds different values of y, then the function is said to be multiple valued function. For example y = 2x2-x + 2 is single-valued function and $y = \pm\sqrt{3x^2 - 5}$ is multiple-valued function.

(iii) *Constant Function :* A function is said to be constant function if for each and every value of x the value of y remains same.

Example: $y = kx^o$.

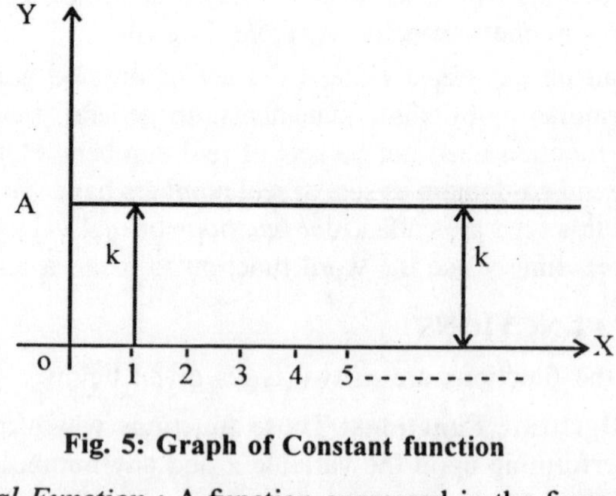

Fig. 5: Graph of Constant function

(iv) *Polymial Function :* A function expressed in the form P(x) = anxn

$+ a_{n-1} x^{n-1} + a_{n-2} x^{n-2} + \text{- - - - - - -} + a_o$. where ao, a1, a2, - - - - an, are known as coefficients with an $\neq$ o, n is positive integer. Then p(x) is called polynomical of degree n in x. The polynomical is defined for all real values of x.

For example

$p(x) = -5x^4 - 3x^2 + 7x + 6$

$p(x) = \sqrt{3}\, x^2 - 2x + 1$

For n = 2, the polynomical of second degree gives Quadratic quation of the form $ax^2 + bx + c = o$ and the zeros of the euation are given by.

where $b^2 - 4$ ac is called discriminant when b^2-4ac>o, the roots (or zeros) of the equation be real and if b^2-4ac < o, the roots be imaginary (or complex) and when b^2-4ac = o, the roots will be equal.

Similarly, an algebraic equation (or polynomial equation) we main an equation of th form $a_n z^n + a_{n-1} + a_{n-2} Z^{n-2} + + \ldots\ldots + a_1 z + a_o =$ o. where n is a non-negative integer, an $\neq$ o $a_o, a_1, a_2, \ldots\ldots$. an are fixed real numbers and z is the variable complex. Then this is called a complex polynomial of degree n.

(v) *Rational Function* : A function R(x) defined in the form R(x) = , where Q(x) ¹ o is said to be rational function. For example R(x) = where p(x) and Q(x) both are polynomials.

(vi) *Explicit and Implicit Function* : If the relation between x and y is given such that the value of y can be easily expressed in term of x, then y is known as explicit function. But if the value of y cannot be expressed in terms of x easily, then y is known as implicit function.

For example $Y^2 - 4x^2 = o$ and $Y = \sqrt{3}\, x$ are explicit function while $x^3 + y^3 + 3xy = o$, and $ax^2 + 2hxy + by^2 = o$ are implicit functions.

(vii) *Modules Function* : A function f is said to be modulus or absolute value function if *f*(x) = /x/

where /x/ = x when $x \geq o$

= -x when x < o

(viii) *Signum Function* : The function f defined as

f(x) = if x ¹

= o if x = o

is called the signum function

Here

$f(x) = 1$ if $x > o$

$= o$ if $x = o$

$= -1$ if $x < o$

Domain of f(x) is the set of all real numbers and the range of the function is the set {-1, o, 1} i.e. {-1, 1}.

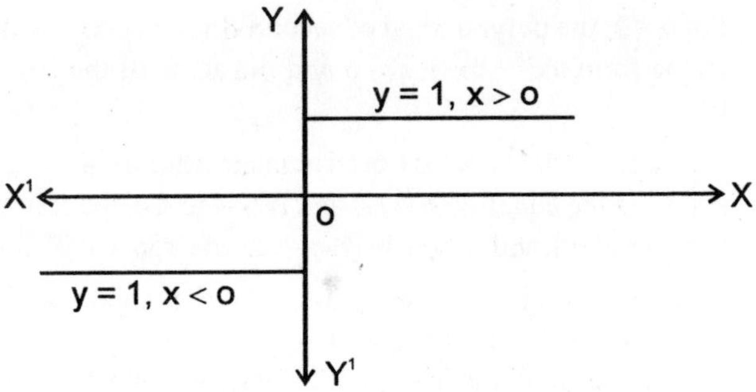

Fig. 6: Graph of Signum Function

(ix) *Power Function :* The function of the form y = x2, , etc are known as Power function.

SOME OTHER FUNCTIONS

(i) *Equal Functions :* Two functions ¦(x) and g(x) are said to be equal, if they have same domain and the range, and ¦(x) = g(x) and x in the domain.

(ii) *Even and odd Functions :* A function is said to be even function if after changing the variable x with (-x) there is no change in the function i.e. $f(-x) = + f(x)$. Similarly, After replacing x by (-x), if the function change by negative, then the function is said to be odd function i.e. $f(-x) = -(fx)$. For example, Sinx, x^3, x^5 are odd functions where cosx, x^2, x^4, are even functions of x.

(iii) *Inverse Functions :* If $y = f(x)$ is any function of x then $x = f^{-1}(y)$ is known as inverse function. Suppose $y = x^2$ is any function then $x = f^{-1}(y) = \pm \sqrt{y}$ is inverse function of x. Similarly, if y = sin x, y = tan x, $y = e^x$ etc are the functions of x then $x = sm^{-1} y$, $x = tan^{-1} y$, x = log y are the inverse functions respectively.

SOLVED EXAMPLES

Example 1:

Describe the function f on A = {-1,0, 1, 2} defined by:

$$f(x) = x^2 + 1$$

Determine the range B of the function f. Is f one-one? Is the inverse relation a function on B?

Solution:

$$f = \{(-1, 2), (0, 1), (1, 2), (2, 5)\}$$

Range of the function = $B = \{2, 1, 5\}$

f is not one-one since, –1 and 1 has the same image.

$$f^{-1} = \{(2, -1), (1, 0), (2, 1), (5, 2)\},$$

which is not a function since 2 has two images.

Example 2:

Consider two sets:

$$A = \{a, b, c\}, \text{ and } B = \{p, q, r, s, t\}$$

Three relations from A to B are defined as:

i. $f = \{(a, p), (b, p), (c, p), (a, r)\}$

ii. $g = \{(a, p), (c, q)\}$

iii. $\phi = \{(a, t), (b, t), (c, t)\}$

State which of them are functions.

Solution:

i. f is not a function since $a \in A$ has no unique image.

ii. g is not a function since $b \in A$ has no image.

iii. ϕ is a function since every element of A has unique image.

Example 3:

If f and g are two functions from to defined as:

$$f(x) = x^2 \text{ and } g(x) = 3x,$$

find fo f, fo g, go g and go f.

Solution:

$$(fo\, f)(x) = f(f(x)) = f(x^2) = (x^2)^2 = x^4$$
$$(fo\, g)(x) = f(g(x)) = f(3x) = (3x)^2 = 9x^2$$
$$(go\, g)(x) = g(g(x)) = g(3x) = 3.3x = 9x$$
$$(go\, f)(x) = g(f(x)) = g(x^2) = 3x^2$$

[Note that $fo\, g \neq go\, f$, if $f \neq g$]

Example 4:

If $f(x) = x^2 - 2x + 3$, show that :

$$f(x + 1) + f(x - 1) = 2\,[1 + f(x)]$$

Solution:

$$\begin{aligned} f(x + 1) &= (x + 1)^2 - 2(x + 1) + 3 \\ &= x^2 + 2x + 1 - 2x - 1 + 3 \\ &= x^2 + 3 \\ f(x - 1) &= (x - 1)^2 - 2(x - 1) + 3 \\ &= x^2 - 2x + 1 - 2x + 1 + 3 \\ &= x^2 - 4x + 5 \end{aligned}$$

$$\therefore \quad \begin{aligned} f(x + 1) + f(x - 1) &= x^2 + 3 + x^2 - 4x + 5 \\ &= 2x^2 - 4x + 8 \\ &= 2\,(x^2 - 2x + 4) \\ &= 2\,(x^2 - 2x + 3 + 1) \\ &= 2\,[f(x) + 1] \end{aligned}$$

Example 5:

Let $2p + 3q = 54$, where p is the price and q the quantity. Indicate the mathematical and meaningful domain and range of this function when:

i. Price is the independent variable

ii. Quantity is the independent variable

Solution:

i. $2p + 3q = 54 \Rightarrow q = \frac{1}{3}(54 - 2p)$

For getting quantity as a positive rational number, we have to get $q > 0$

$\Rightarrow \frac{1}{3}(54 - 2p) > 0$

$\Rightarrow 54 - 2p > 0$

$\Rightarrow 54 > 2p$, or, $p < 27$

Hence,

domain = $\{p : 0 < p < 27, p \text{ is a rational number}\}$, and

range = $\{q : 0 < q < 18\ , q \text{ is a rational number}\}$

ii. $2p + 3q = 54 \Rightarrow p = (54 - 3q)$ For getting price as a positive rational number, we have to get $p > 0$

$\Rightarrow \frac{1}{2}(54 - 3q) > 0$

$\Rightarrow 54 - 3q > 0$

$\Rightarrow 54 > 3q$, or, $q < 18$

Hence,

Domain = $\{q : 0 < q < 18$, q is a rational number$\}$, and

Range = $\{p : 0 < p < 27$, p is a rational number$\}$.

Example 6:

Find which of the following functions are one-one and which of them are many one, where:

i. $f = \{(1, 3), (2, 2), (3, 1), (4, 4)\}$

ii. $\phi(p) = 1, \phi(q) = 1, \phi(r) = 10, \phi(s) = 100$

iii. $\psi(x) = x^2, x \in \mathbb{R}$ [$\mathbb{R}$ is the set of real numbers].

iv. $\theta(x) = 5x, x \in$ I [I is the set of integers].

Solution:

i. f is one-one, since,
every distinct element of the domain has unique image.

ii. ϕ is many one, since, both the elements p and q of the domain has same image.

iii. Here we get ψ as $\{(1, 1), (-1, 1), (2, 4), (-2, 4), \ldots\}$.
Hence ψ is many one.

iv. Here we get θ as $\{(1,5), (-1,-5), (2,10), (-2,-10), \ldots\}$.
Hence θ is one-one.

Example 7:

Find which of the following functions from A to A are into and which of them are onto, where:

i. $A = \{1, 2, 3, 4\}, f_1 = \{(1, 3), (2, 3), (3, 2), (4, 2)\}$

ii. $A = \{x, y, z\}, f_2 = \{(x, y), \{(y, z), \{(z, x)\}$

iii. $A = \{1, -1\}, f_3 (x) = x^2$

iv. $A = \{1, -1\}, f_4 (x) = x^3$

Solution:

i. f_1 is into ,since
range of $f_1 = \{2, 3\}$ is a proper subset of A

ii. f_2 is onto , since
range of $f_2 = \{x, y, z\} = A$

iii. Here we get f_3 as {(1, 1), (-1, 1)}.
Hence f_3 is into.

iv. Here we get f_4 as {(1, 1), (-1, -1)}.
Hence f_4 is onto.

EXERCISES

1. Prove that $z_1 . z_2 = o$ implies $z_1 = o$ or $z_2 = o$
2. Show that, although for real numbers $x^2 + y^2 = o$ implies $x = o$ and $y = o$, for Complex number $z_1^2 + z_2^2 = o$ does not imply $z_1 = o$ or $z_2 = o$.
3. If $f(x) = \cos(\log x)$, then evaluate $f(x), f(y) - 1/2 [f(x/y)$
4. If the function is given by $f(x) = 1/2\ (a^x + a^{-x})$, $a > o$, show that
$f(x + y) + f(x - y) = 2f(x) f(y)$.
5. The total cost C of a firm per day is a function of its daily output q given by $C = 170 + 8q$. The firm has a capacity limit of 200 units of output per day. What are the domain and range of the cost function C?

$$f(x) = \begin{cases} \frac{1}{2} - x & 0 \le x < \frac{1}{2} \\ 1 & x = \frac{1}{2} \\ \frac{3}{2} - x & \frac{1}{2} < x \le 1 \end{cases}$$

6. Show that $f(x)$ is continuous at $x = 1$, if

$$f(x) = \begin{cases} 5x - 4 & 0 < x \le 1 \\ 4x^3 - 3x & 1 < x < 2 \end{cases}$$

7. Let $f : R \rightarrow$ be given by $f(x) = \text{Sin } x$ and
$g : R \rightarrow$ be given by $g(x) = x^2$, find $(go f)$ and $(f og)$.

2

Limits and Continuity

INTRODUCTION

Consider the function, $f(x) = 2x + 3$. As x assumes different values, the value of $f(x)$ will also change. When $x = 5$, $f(x) = 13$, when $x = 0$, $f(x) = 3$; and when $x = -2$, $f(x) = -1$.

For this function $f(x)$, it is possible to find its value for any finite value of its variable x. In general, let x be a real variable and a be a finite real number. We denote the value of the function $f(x)$ when x is replaced by a as $f(a)$.

Hence in $f(x) = 2x + 3$, we write:

$$f(5) = 13; f(0) = 3 \text{ and } f(-2) = -1$$

But for some functions $f(x)$, on putting $x = a$, it will not give a defined number. For example, let us find the value of the function:

$$f(x) = \frac{x^2 - 1}{x - 1} \text{ at } x = 1$$

We get $f(1) = 0/0$, which is an indeterminate quantity. But for all other finite real values of x (except 1), we get defined values for the function. Consider another function:

$$f(x) = \frac{1}{x - 5}$$

Its value at $x = 5$ is $1/0$, which is undefined as division by zero is not admissible. Let us see one more function, say,

$$f(x) = \frac{1}{\sqrt{x - 2}}$$

For this function, $f(2)$ is undefined as in the previous case. But, what is the case with $f(1)$? We get $f(1) = \frac{1}{\sqrt{-1}}$, which is also undefined as it contains root of a negative quantity. i.e., this $f(x)$ is defined only for numbers greater than 2. Thus, there are functions that seize to be defined in certain points. In such cases it is possible to find the value $f(a)$ to where $f(x)$ 'approaches' as x approaches a. This means that as x comes closer and closer to 1, $f(x)$ comes closer and closer to 2. We say the 'limit value' of $f(x) = \frac{x^2-1}{x-1}$ at $x = 1$ is 2. This is written as:

$$\lim_{x \to 1} f(x) = 2$$

That is, *limit* of a function is that fixed value to which a function approaches as the variable approaches a given value. More formally, we define *the limit of a function f(x) when $x \to a$ as l, if corresponding to a positive number ε however small we can find a number δ such that $|f(x)-l| < \varepsilon$, for all values of x satisfying $|x-a| < \delta$.* [This definition is known as the ε -δ definition of the limit of a function]. According to this definition, l is the limit of $f(x)$ at $x = a$, if the distance between $f(x)$ and l can be made as small as we please for all values of x for which the distance between x and a is sufficiently small.

We write this as:

$$\lim_{x \to a} f(x) = l$$

PROPERTIES OF LIMITS

1. $\lim_{x \to a} [f(x) \pm \phi(x)] = \lim_{x \to a} f(x) \pm \lim_{x \to a} \phi(x)$

2. $\lim_{x \to a} k.\, f(x) = k.\, \lim_{x \to a} f(x)$, k being a constant

3. $\lim_{x \to a} [f(x) \times \phi(x)] = \lim_{x \to a} f(x) \times \lim_{x \to a} \text{f}(x)$

4. $\lim_{x \to a} \frac{f(x)}{g(x)} = \frac{\lim_{x \to a} f(x)}{\lim_{x \to a} g(x)}$, if $\lim_{x \to a} g(x) \neq 0$

5. $\lim_{x \to a} \frac{1}{f(x)} = \frac{1}{\lim_{x \to a} f(x)}$, if $\lim_{x \to a} f(x) \neq 0$

6. $\lim_{x \to a} \log f(x) = \log \lim_{x \to a} f(x)$, if $\lim_{x \to a} f(x) > 0$

SOME IMPORTANT LIMITS

The following are some limits which we want to remember, since they are frequently used in problems. However we skip the proof.

i. $\lim_{x \to a} \dfrac{x^n - a^n}{x - a} = n\, a^{n-1}$, for any natural number n

ii. $\lim_{x \to a} \dfrac{\sin\theta}{\theta} = 1$, θ is measured in radians

iii. $\lim_{x \to a} \dfrac{e^x - 1}{x} = 1$

iv. $\lim_{x \to a} \dfrac{\log(1+x)}{x} = 1$

v. $\lim_{x \to a} \dfrac{a^x - 1}{x} = \log a$

To evaluate limits, there are many methods such as direct substitution, factorization or rationalization. Some limits can be evaluated using known limit formulae. [Another method of evaluating limits using L'Hospital's Rule is given in Chapter-9]

Example:

Evaluate the following:

i. $\lim_{x \to -2} (x^2 - 2x + 5)$

ii. $\lim_{x \to 0} (1 - \text{Sin } 2x)$

Solution

i. $\lim_{x \to -2} (x^2 - 2x + 5) = (-2)^2 - 2(-2) + 5$

$= 4 + 4 + 5 = 13$

ii. $\lim_{x \to 0} (1 - \text{Sin } 2x) = 1 - \text{Sin } 0$

$= 1 - 0 = 1$

LEFT HAND AND RIGHT HAND LIMITS

It is clear that as x increases and comes closer to 1 from left hand side of 1, the value of $f(x)$ also increases and comes closer to 2. This means that the left hand limit of $f(x)$ at $x = 1$ is 2. We use '$x \to 1^-$' to denote x approaches 1 from the left hand side. Hence it can be written as:

$$\lim_{x \to 1^-} f(x) = 2$$

Also from the above tables it is obvious that as x decreases and comes closer to 1 from right hand side of 1, the value of $f(x)$ also decreases and comes closer to 2. This means that the right hand limit of $f(x)$ at $x = 1$ is 2. We use '$x \to 1^+$' to denote x approaches 1 from the right hand side; and we write:

$$\lim_{x \to 1^+} f(x) = 2$$

Here we get,

$$\lim_{x \to 1^-} f(x) = \lim_{x \to 1^+} f(x)$$

i.e., in this case, the left hand limit (LHL) = the right hand limit (RHL).

There are cases where the left and right limits do not coincide. Consider a function defined by:

$$f(x) = \frac{|x|}{x},\ x \neq 0$$

Here $f(x)$ is defined for all values of x except for $x = 0$. At $x = 0$, the value of the function turns indeterminate.

Example: Evaluate the following:

i. $\lim_{x \to -2} (x^2 - 2x + 5)$

ii. $\lim_{x \to 0} (1 - \text{Sin } 2x)$

iii. $\lim_{x \to 2} \frac{\sqrt{6-x} - \sqrt{6+x}}{x}$

iv. 1.5 Definition Limit

A function f(x) is said to tend to the limit A as x tends to a, if corresponding to any arbitrary chosen positive number ε, however small (but not zero), there exists a positive number δ, such that

$$|f(x) - A| < \varepsilon,$$

for all values of x for which $0 < |x - a| < \delta$.

We write it as $\lim_{x \to a} f(x) = A$.

Right Hand Limit: A function f(x) is said to tend to the limit A as $x \to a$ from the right, if corresponding to any arbitrary chosen +ive number ε, however small (but not zero), there exists a +ive number δ, such that

$$|f(x) - A| < \varepsilon,$$

for all values of x for which $0 < x - a < \delta$.

The limit of f(x) as $x \to a$ from the right is called the *right hand limit* of f(x) and is denoted by $\lim_{x\to a+0} f(x)$ or by $f(a + 0)$.

We calculate $f(a + 0)$ by evaluating $\lim_{h\to 0} f(a + h)$, where h is +ive and sufficiently small.

Left Hand Limit: A function f(x) is said to tend to the limit A as $x \to a$ from the left, if corresponding to any arbitrary chosen +ive number ε, however small (but not zero), there exists a +ive number δ, such that

$$|f(x) - A| < \varepsilon,$$

for all values of x for which $0 < a - x < \delta$.

The limit of f(x) as $x \to a$ from the left is called the *left hand limit* of f(x) and is denoted by $\lim_{x\to a-0} f(x)$ or by $f(a + 0)$.

We calculate $f(a - 0)$ by evaluating $\lim_{h\to 0} f(a - h)$, where h is +ive and sufficiently small.

From the definition it is clear that the limit of f(x) as $x \to a$ has no necessary connection with the value of the function at a except that when the function is continuous at x = a: The two are equals; even in the case of such functions the limit is distinct in idea from the value.

Note 1: $\lim_{x\to a} f_1(x)$ exists only if, limit on the left = limit on the right.

Note 2: If $f(x) \to A$ and $f_2(x) \to B$, as $x \to a$, where A and B are both finite, we have

(i) $\lim_{x\to a} \{f_1(x) \pm f_2(x)\} = A \pm B$,

(ii) $\lim_{x\to a} \{f_1(x) \,.\, f_2(x)\} = A \,.\, B$,

(iii) $\lim_{x\to a} \left\{\dfrac{f_1(x)}{f_2(x)}\right\} = \dfrac{A}{B}$, provided $B \neq 0$,

(iv) $\lim_{x\to a} cf(x) = cA$ where c is constant.

LIMIT OF INEQUALITIES

It $\phi(x) + A$ and $\psi(x) \to B$ as $x \to a$ and if

$\phi(x) < \psi(x),\ x \neq u$

Then $A \leq B$

Since $\phi(x) < \psi(x)$ for all values of a an which the limit depend one would naturally expect that A < B. But it is also possible that A may be equal to B. Let $\phi(x) = 1 + x^2$. and $\psi(x) = 1 + 2u^2$. We have $\lim\limits_{x \to a} \phi(x) = 1$, $\lim\limits_{x \to a} \psi(x) = 1$. Also if $x \neq 0$ we have $\phi(x) < \psi(x)$ for all values of x on which the limit depend.

DIFFERENTIABILITY

Definition: *A function f(x) is said to be differentiable at x = a if both*

$$\lim_{h \to 0} \frac{f(a+h) - f(a)}{h}, h > 0 \text{ and } \lim_{h \to 0} \frac{f(a-h) - f(a)}{-h}, \text{ (L.H.L. = R.H.L.)}$$

exist and have a common value (finite or infinite). This common value is called the *derivative* of f(x) at the point x = a and is denoted by f'(a). Another name for derivative is known as differential co-efficient.

Progressive and Regressive Derivatives: Progressive derivative or *the right hand differential coefficient of* f(x) at x = a is given by

$$\lim_{h \to 0} \frac{f(a+h) - f(a)}{h}, h > 0.$$

It is denoted by Rf'(a).

Regressive derivative or *the left hand differential coefficient* of f(x) at x = a is given by

$$\lim_{h \to 0} \frac{f(a-h) - f(a)}{-h}, h > 0.$$

It is denoted by Lf'(a).

A function f(x) is said to be differentiable at x = a if both Rf'(a) and Lf '(a) exist and are equal; otherwise it is said to be not differential.

Theorem:

Continuity is necessary but not a sufficient condition for the existence of a finite derivative.

Proof:

Let f(x) have a finite derivative f'(a) at x = a. Then

$$\lim_{h \to 0} \frac{f(a+h) - f(a)}{h} = f'(a). \quad ...(1)$$

$$\text{Now } f(a + h) - f(a) = \frac{f(a-h) - f(a)}{h} \times h, h \neq 0.$$

Taking limit of both sides when h → 0, we get

$$\lim_{h\to 0}\left[f(a+h)-f(a)\right]=\lim_{h\to 0}\frac{f(a+h)-f(a)}{h}\times\lim_{h\to 0}h$$

$= f'(a) \times 0,$ [from (1)]

$= 0$, since f'(a) is finite.

Now, $\lim_{h\to 0}$ [f(a + h) – f (a)] = 0

$\Rightarrow$ $\lim_{h\to 0}$ f(a + h) = f(a). ...(2)

From the condition (2) we see that f(x) is necessarily continuous at x = a. Thus continuity is a necessary condition for the existence of a finite derivative.

In order to show that continuity is not a sufficient condition for differentiability, we should give an example of a function which is continuous at a point but is not differentiable at that point.

CONTINUITY

Definition: *A function f(x) defined for x = a is said to be continuous at x = a if*

(i) f(a) i.e., the value of f(x) at x = a is a definite number.

(ii) the limit of the function f(x) as x → a exists and is equal to the value of f(x) at x = a.

Otherwise the function is *discontinuous* at x = a.

A function is said to be continuous in an interval (a, b) if it is continuous at every point of that interval.

Arithmetical Definition of Continuity: *A function f(x) is said to be continuous at x = a, if for any arbitrarily chosen positive number ε, however small (but not zero), we can find a corresponding number δ such that*

$$|f(x) - f(a)| < \varepsilon$$

for all values of x for which $|x - a| < \delta$.

This definition, as given by Cauchy, is known as Cauchy's definition of continuity.

Remark:

One comparing the definitions of limit and continuity, we find that a function f(x) is continuous at x = a if

$$\lim_{x\to a} f(x) = f(a).$$

Thus f(x) is continuous at x = a if we have L.H.L. = R.H.L. = f(a); otherwise it is discontinuous at x = a.

ELEMENTARY FUNCTION OF CONTINUITY

The student as supposed to be familiar with the elementary functions and to know that:

(i) x^n is continuous for all values of x when n is positive.

(ii) x^n is continuous for all values of x except x = 0 when n is negative. For when n is negative and equal to –m say, where m iis positive we can write x^n as $\frac{1}{x^m}$ and as $x \to 0$ $\frac{1}{x^m}$ either does not tends to a limit or $\to \infty$.

(iii) sin x and cos x are continuous for all values of x.

(iv) tan x is continuous for all values of x except $\pm\frac{\pi}{2}, \pm\frac{3\pi}{2}, \pm\frac{5\pi}{2}$ etc. as $x \to \frac{\pi}{2}$ from the right tan $x \to -\infty$ as $x \to \frac{\pi}{2}$ from the left tan $x \to +\infty$. Thus, $\lim_{x\to\pi/2}$ tan x does not exist.

(v) Similarly, statement can be made for cot x, sec x and cosec x.

(vi) $\sin^{-1}x$ and $\cos^{-1}x$ are not real for value of x out side the domain [– 1, 1] similarly, $\sec^{-1} x$ and $\text{cosec}^{-1} x$ are not real for values of x between + and.

(vii) All the inverse circular function are continuous at every point with in the domain in which they are real.

(viii) Log x is continuous for all values of x > 0. For values of x < 0 the function is not real and we shall not consider these values as x tends to zero from the right log x tends to $-\infty$.

CONTINUITY OF SUM PRODUCIS

(i) The sum of two continuous function is continuous.

(ii) The product of two continuous function is continuous.

(iii) The quotient of two continuous function is a continuous function except for value of x which make the denominator zero.

SOLVED EXAMPLES

Example 1:

Evaluate: $\lim_{x\to 0}\frac{1}{1-e^{1/x}}$.

Solution:

We have $\lim_{x\to 0}\frac{1}{1-e^{1/x}}$

Let $f(x) = \dfrac{1}{1 - e^2x}$

We have $f(0 + 0) = \lim_{h\to 0} f(h) = \lim_{h\to 0} \dfrac{1}{1 - e^{1/h}} = 0$

Again $f(0 - 0) = \lim_{h\to 0} f(-h) = \lim_{h\to 0} \dfrac{1}{1 - e^{-1/h}}$

$$= \lim_{h\to 0} \frac{1}{1 - (1/e^{1/h})} = 1$$

L.H.L $\neq$ R.H.L

$\because$ $f(0 + 0) \neq f(0 - 0)$ $\therefore$ $\lim_{x\to 0} \dfrac{1}{1 - e^{1/x}}$ does not exist.

Example 2:

Evaluate: $\lim_{x\to 0} \dfrac{\sin x}{x}$

Solution:

Let $f(x) \dfrac{\sin x}{x}$

We have R.H.L. $= f(0 + 0) = \lim_{h\to 0} f(0 + h) = \lim_{h\to 0} f(h) = \lim_{h\to 0} \dfrac{\sin h}{h}$

$$= \lim_{h\to 0} \frac{h - \frac{h^3}{3!} + \frac{h^5}{5!} - \ldots}{h} = \lim_{h\to 0} \left(1 - \frac{h^2}{3!} + \frac{h^4}{5!} - \ldots\right) = 1$$

L.H.L., $= f(0 - 0) = \lim_{h\to 0} f(0 - h) = \lim_{h\to 0} f(-h)$

$$= \lim_{h\to 0} \frac{\sin(-h)}{h} = \lim_{h\to 0} \frac{\sin h}{h} = 1.$$

L.H.L. = R.H.L.

Since $f(0 + 0) \neq f(0 - 0) = 1$ therefore $\lim_{x\to 0} \dfrac{\sin x}{x} = 1.$

Example 3:

The limit of a sum is equal to the sum of the limits.

Solution:

Let $\lim_{x\to a} f_1(x) = A$ and $\lim_{x\to a} f_2(x) = B$.

Let ε be any arbitrary chosen positive number.

We can choose δ_1, δ_2 such that

$$|f_1(x) - A| < \frac{1}{2}\varepsilon, \text{ when } 0 < |x - a| < \delta_1,$$

and $$|f_2(x) - B| < \frac{1}{2}\varepsilon, \text{ when } 0 < |x - a| < \delta_2,$$

Let δ be any positive number smaller than both δ_1 and δ_2. Then

$$|f_1(x) - A| < \frac{1}{2}\varepsilon \text{ and } |f_2(x) - B| < \frac{1}{2}\varepsilon, \text{ for } 0 < |x - a| < \delta$$

Now from algebra, we have

$$|\{f_1(x) + f_2(x)\} - (A + B)| = |\{f_1(x) - A\} + f_2(x)\} - B)\}|$$

$$\leq |\{f_1(x) - A| + |f_2(x)\} - B)\}| < \frac{\varepsilon}{2} + \frac{\varepsilon}{2} \textit{ i.e., } < \varepsilon$$

$\therefore$ $$\lim_{x\to a}\{f_1(x) + f_2(x)\} = A + B.$$

Example 4:

The limit of a product is equal to the product of the limits.

Solution:

Let $\lim_{x\to a} f_1(x) = A$ and $\lim_{x\to a} f_2(x) = B$.

Let $A \neq 0$ and $B \neq 0$

Again let, $f_1(x)f_2(x) - AB$

$$= [f_1(x) - A]\,[f_2(x)\} - B] + A\,[f_2(x) - B] + B[f_1(x) - A].$$

$\therefore$ by algebra, we have $|f_1(x) - f\,|\,f_2(x) - AB|$

$$\leq |f_1(x) - A|\,.\,|f_2(x) - B)| + |A|\,.\,|f_2(x) - B| + |B|\,f_1(x) - A|$$

Now, $\lim_{x\to a} f_1(x) = A$ and $\lim_{x\to a} f_2(x) = B$. Therefore, for a given ε we can take δ such that

$$|f_1(x) - A| < \frac{\varepsilon}{3|B|} \text{ and } |f_2(x) - B| < \frac{\varepsilon}{3|A|},$$

when $0 < |x - a| < \delta$. Then

$$|f_1(x)f_2(x) - AB| < \frac{\varepsilon}{3}\cdot\frac{\varepsilon}{3|A||B|} + \frac{\varepsilon}{3} + \frac{\varepsilon}{3}.$$

Since only small values of ε need by considered, we can take ε such that $\frac{\varepsilon}{3|A||B|} < 1$. Then

$$|f_1(x)f_2(x) - AB| < \frac{\varepsilon}{3} + \frac{\varepsilon}{3} + \frac{\varepsilon}{3} \text{ i.e, } < \varepsilon, \text{ when } 0 < |x - a| < \delta.$$

Hence $\lim_{x\to a}\{f_1(x)\,.\,f_2(x)\} = AB.$

Example 5:

Evaluate: $\lim_{x\to\infty}\dfrac{\sin x}{x}$.

Solution:

Put $x = \dfrac{1}{y}$. Then $= \lim_{h\to\infty} = \dfrac{\sin x}{x} = \lim_{y\to 0} y \sin\left(\dfrac{1}{y}\right)$

Now let $f(y) = y \sin\left(\dfrac{1}{y}\right)$.

We have R.H.L. $f(0 + 0) = \lim_{h\to 0} f(0 + h) = \lim_{h\to 0} f(h) = \lim_{h\to 0} h \sin\left(\dfrac{1}{h}\right) = 0$

Since $\sin\left(\dfrac{1}{h}\right)$ lies between –1 and 1.

Also L.H.L. $f(0 - 0) = \lim_{h\to 0} f(0 - h) = \lim_{h\to 0} f(-h) = \lim_{h\to 0} (-h)$

$$= \lim_{h\to 0} f \sin\left(-\frac{1}{h}\right) = \lim_{h\to 0} h \sin\left(\frac{1}{h}\right) = 0$$

Since both the limits $f(0 + 0)$ and $f(0 - 0)$ exist and are equal to zero, therefore $\lim_{y\to 0} y \sin\left(\dfrac{1}{y}\right) = 0$, Hence $\lim_{x\to\infty}\dfrac{\sin x}{x} = 0$.

Example 6:

Evaluate: $\lim_{x\to 0} = \sin\dfrac{1}{x}$.

Solution:

Let $f(x) = \sin\dfrac{1}{x}$

We have $f(0 + 0) = \lim_{h\to 0} f(0 + h) = \lim_{h\to 0} f(h) = \lim_{h\to 0} \sin\left(\dfrac{1}{h}\right)$

As $h \to 0$, the value of $\sin\left(\dfrac{1}{h}\right)$ oscillates between + 1 and –1, passing through zero and intermediate values an infinite number of times. Thus, there is no definite number A to which $\sin\left(\dfrac{1}{h}\right)$ tends as h hends to zero. Therefore the right hand limit $f(0 + 0)$ does not exist.

Similarly, the left hand limit f(0 – 0) also does not exist. Hence $\lim_{x\to 0} \sin\frac{1}{x}$ does not exist.

Example 7(a):

Evaluate: $\lim_{x\to a} = \frac{x^m - a^m}{x - a}$.

Solution:

$$\lim_{x\to a} \frac{x^m - a^m}{x - a}$$

Let $f(x) = \frac{x^m - a^m}{x - a}$

We have R.H.L. $f(a + 0) = \lim_{h\to 0} f(a + h) = \lim_{h\to 0} = \frac{(a-h)^m - a^m}{a + h - a}$

$$= \lim_{h\to 0} \frac{a^m\left[\left(1+\frac{h}{a}\right)^m - 1\right]}{h}$$

$$= \lim_{h\to 0} \frac{a^m}{h}\left[1 + m.\frac{h}{a} + \frac{m(m-1)}{2!}\frac{h^2}{a^2} + \ldots - 1\right]$$

L.H.L. $= f(a - 0) = \lim_{h\to 0} f(a - h) = \lim_{h\to 0} \frac{(a-h)^m}{a - h - a} = ma^{m-1}$.

Left hand limit = Right hand limit

Thus, $f(a + 0) = f(a - 0) = ma^{m-1}$

Therefore $\lim_{x\to a} f(x) = ma^{m-1}$.

Example 7(b):

Evaluate: $\lim_{x\to 0} \frac{1}{x} e^{1/x}$

Solution:

Let $f(x) = \frac{1}{x} e^{1/x}$

R.H.L. $= f(0 + 0) = \lim_{h\to 0} f(0 + h) = \lim_{h\to 0} f(h) = \lim_{h\to 0} \frac{1}{h} e^{1/h} = \infty$

[Since both and $e^{1/h}$ tend to ∞ as $h\to 0$]

Again we have

$$\text{L.H.L.} = f(0-0) = \lim_{h\to 0} f(0-h) = \lim_{h\to 0} f(-h) = \lim_{h\to 0} -\frac{1}{h}e^{-1/h}$$

$$= \lim_{h\to 0}\frac{-1}{he^{1/h}} = \lim_{h\to 0}\frac{-1}{h\left(1+\frac{1}{h}+\frac{1}{2!}\frac{1}{h^2}+...\right)}$$

$$= \lim_{h\to 0}\frac{-1}{h+1+(1/2h)+...} = 0.$$

L.H.L. ≠ R.H.L. Since both the limits f(0 + 0) and f(0 – 0) are unequal, therefore $\lim_{x\to 0}\frac{1}{x}e^{1/x}$ does not exit.

Example 8:

Here the right hand limit i.e., f(0 + 0) = h → 0 f(0 – h), where h is +ive and sufficiently small

Evaluate: $\lim_{x\to 0}\frac{e^x-1}{x}$.

Solution:

$= \lim_{h\to 0} f(h) = \lim_{h\to 0} h$, [∵ $h > 0$ and $f(x) = x$ when $x > 0$]

$= 0.$

Again the left hand limit i.e., f(0 – 0) = $= \lim_{h\to 0} f(0-h)$,

where h is +ive and sufficiently small

$= \lim_{h\to 0} f(-h) = \lim_{h\to 0} -(-h)$, [∵ $-h < 0$ and $f(x) = -x$ when $x < 0$]

$= \lim_{h\to 0} h = 0.$

Since both the limits f(0 + 0) and f(0 – 0) exist and are equal to zero, therefore $\lim_{x\to 0} f(x)$ *exists and is equal to zero.*

Example 9:

Thus, f(0 + 0) = f(0 – 0) = 1. Therefore $\lim_{x\to 0}$ $(e^x - 1)/x = 1$.

Evaluate: $\lim_{x\to 0}\frac{a^x-1}{x}$.

Solution:

$$\lim_{x\to 0}\frac{a^x-1}{x}.\text{Let } f(x) = \frac{a^x-1}{x}$$

$$= \frac{1 + x \log a + \left(x^2/2!\right)(\log a)^2 + ... - 1}{x}$$

Now, $= \dfrac{x\left[\log a + \frac{1}{2} x (\log a)^2 + ...\right]}{x}$

R.H.L. $= f(0+0) = \lim_{h \to 0} f(h) = \lim_{h \to 0} \dfrac{h\left[\log a + \frac{h}{2}(\log a)^2 + ...\right]}{h} = \log a$.

L.H.L. $= f(0 - 0) = \lim_{h \to 0} f(0 - h) = \lim_{h \to 0} f(-h) = \log a.$

Thus, $f(0 + 0) = f(0 - 0) = \log a$. Therefore

$$\lim_{x \to 0} \frac{\left(a^x - 1\right)}{x} = \log a \cdot$$

Example 10(a):

Find the right hand and the left hand limits in the following cases and discuss the existence of the limit in each case:

(i) $\lim_{x \to 2} \dfrac{2x^2 - 8}{x - 2}$;

(ii) $\lim_{x \to 0} \dfrac{e^{1/x} - 1}{e^{1/x} + 1}$;

(iii) $\lim_{x \to 0} f(x)$, *where f(x) is defined as*

$f(x) = x$, *when* $x > 0$

$f(x) = 0$, *when* $x = 0$

$f(x) = -x$, *when* $x < 0$

Solution:

Let $f(x) = \dfrac{2x^2 - 8}{x - 2}$.

We have L.H.L. $= f(2+0) = \lim_{h \to 0} f(2+h) = \lim_{h \to 0} \dfrac{2(2+h)^2 - 8}{2 + h - 2}$

$$= \lim_{h \to 0} \frac{2\left(4 + 4h + h^2\right) - 8}{h} = \lim_{h \to 0} \frac{8h + 2h^2}{h} = \lim_{h \to 0} \frac{h(8 + 2h)}{h}$$

$$= \lim_{h \to 0} (8 + 2h) = 8.$$

Again L.H.L. *i.e.* = $f(2-0)=\lim_{h\to 0} f(2-h)=\lim_{h\to 0}\frac{2(2-h)^2-8}{2-h-2}$

$$=\lim_{h\to 0}\frac{2(4-4h+h^2)}{-h}=\lim_{h\to 0}\frac{-8h+2h^2}{-h}$$

$$=\lim_{h\to 0}\frac{-h(8-2h)}{-h}=\lim_{h\to 0}(8-2h)=8\,.$$

Since f(2 + 0) = f(2 – 0) = 8, therefore $\lim_{x\to 0}\frac{2x^2-8}{x-2}$ exists and is equal to 8.

(ii) Let f(x) $f(x)=\frac{e^{1/x}-1}{e^{1/x}+1}$

We have R.H.L.

$$f(0+0)=\lim_{h\to 0} f(0+h)=\lim_{h\to 0} f(h)=\lim_{h\to 0}\frac{e^{1/h}-1}{e^{1/h}+1}$$

$$=\lim_{h\to 0}\frac{e^{1/h}\left[1-\left(1/e^{1/h}\right)\right]}{e^{1/h}\left[1+\left(1/e^{1/h}\right)\right]}=\lim_{h\to 0}\frac{1-\left(1/e^{1/h}\right)}{1+\left(1/e^{1/h}\right)}=1\,.$$

Again L.H.L. = $f(0-0)=\lim_{h\to 0} f(0-h)=\lim_{h\to 0} f(-h)=\lim_{h\to 0}\frac{e^{-1/h}-1}{e^{-1/h}+1}$

$$=\lim_{h\to 0}\frac{\left(1/e^{1/h}\right)-1}{\left(1/e^{1/h}\right)+1}=\frac{0-1}{0+1}=-1\,.$$

L.H.L. ≠ R.H.L.

Since f(0 + 0) ≠ f(0 – 0), therefore $\lim_{x\to 0}\frac{e^{1/x}-1}{e^{1/x}+1}$ does not exist.

Example 10)b):

Evaluate: $\lim_{x\to 0}\frac{(1+x)^n-1}{x}$.

Solution:

$\lim_{x\to 0}\frac{(1+x)^n-1}{x}$. Let $f(x)=\frac{(1+x)^n-1}{x}$

R.H.L. f(0 + 0) = $\lim_{x\to a} f(h)=\lim_{x\to a}\frac{(1+h)^n-1}{h}$

$$= \lim_{h \to 0} \frac{1 + nh + \frac{n(n-1)}{2!}h^2 + ... - 1}{h}$$

$$= \lim_{h \to 0} \frac{h\left[n + \frac{n(n-1)}{2!}h + ...\right]}{h}$$

$$= \lim_{h \to 0}\left[n + \frac{n(n-1)}{2!}h + ...\right] = n.$$

$$\text{L.H.L.} = f(0-0) = \lim_{h \to 0} f(-h) = \lim_{h \to 0} \frac{(1-h)^n - 1}{-h}$$

$$= \lim_{h \to 0} \frac{1 + n(-h) + \frac{n(n-1)}{2!}(-h)^2 + ... - 1}{-h} = n \cdot$$

R.H.L. = L.H.L.

Thus, f(0 + 0) = f(0 – 0) = n. Therefore, $\lim_{x \to 0} f(x) = n$.

Example 11:

Define the limit of a function. Discuss the existence of the limit of the function f(x) = 1, if x < 1; f(x) = 2 – x, if 1 < x < 2; f(x) = 2, if x ≥ 2 at x = 1 and x = 2.

Solution:

For definition of limit (see definition)

R.H.L. At x = 1,

i.e., $f(1 + 0) = \lim_{h \to 0} f(1 + h)$,

where h is +ive and sufficiently small

$= \lim_{h \to 0} [2 - (1 + h) = \lim_{h \to 0} (1 - h) = 1;$

L.H.L. At x = 1,

i.e., $f(1 - 0) = \lim_{h \to 0} f(1 - h) = \lim_{h \to 0} (1) = 1$

since R.H.L. = L.H.L. = 1,

therefore = $\lim_{x \to 1} f(x)$ exists and is equal to 1.

R.H.L At x = 2,

i.e., $f(2 + 0) = \lim_{h\to 0} f(2 + h) = \lim_{h\to 0} (2) = 2;$

L.H.L. At x = 2,

i.e., $f(2 - 0) = \lim_{h\to 0} (2 - h) = \lim_{h\to 0} [2 - (2 - h)]$

$= \lim_{h\to 0} h = 0.$

$\because$ left hand limit $\neq$ right hand limit,

$\therefore$ $f = \lim_{x\to 2} f(x)$ does not exist.

Example 12(a):

Prove that $\lim_{h\to 0} (1 + x)^{1/x} = e.$

Solution:

Let $f(x) = (1 + x)^{1/x}.$

We have R.H.L.= $f(0 + 0) = \lim_{h\to 0} f(0 + h) = \lim_{h\to 0} f(h) = \lim_{h\to 0} (1 + h)^{1/h}$

$$= \lim_{h\to 0}\left[1 + \frac{1}{h}.h + \frac{\frac{1}{h}\left(\frac{1}{h}-1\right)}{1.2}h^2 + \frac{\frac{1}{h}\left(\frac{1}{h}-1\right)\left(\frac{1}{h}-2\right)}{1.2.3}h^3 + \ldots\right]$$

$$= \lim_{h\to 0}\left[1 + \frac{1}{1!} + \frac{1.(1-h)}{2!} + \frac{1.(1-h)(1-2h)}{3!} + \ldots\right]$$

$$= 1 + \frac{1}{1!} + \frac{1}{2!} + \frac{1}{3!} + \ldots\infty = e.$$

We have, R.H.L. $f(0 - 0) = \lim_{h\to 0} f(0 - h) = \lim_{h\to 0} f(-h)$

$= \lim_{h\to 0} f(1 - h)^{-1/h} = e.$

Since both f(0 + 0) and f(0 – 0) exist and are equal to e, therefore $\lim_{x\to 0} (1 + x)^{1/x} = e.$

Example 12(b):

Evaluate the following limits if they exist: $\lim_{x\to 2} \frac{x^2 - 4}{x - 2}.$

Solution:

Let $f(x) = \frac{x^2 - 4}{x - 2} = \frac{(x-2)(x+2)}{x-2}$

R.H.L. $x = 2,\ f(2 + 0) = \lim_{h\to 0} f(2 + h)$

$$= \lim_{h\to 0} \frac{(2+h-2)(2+h+2)}{2+h-2} = \lim_{h\to 0} \frac{h(h+4)}{h}$$

$$= \lim_{h\to 0} (h + 4) = 4.$$

L.H.L. $x = 2$, *i.e.*, $f(2 - 0) = \lim_{h\to 0} f(2 - h)$

$$= \lim_{h\to 0} \frac{(2-h-2)(2-h+2)}{2-h-2} = \lim_{h\to 0} \frac{-h(-h+4)}{-h}$$

$$= \lim_{h\to 0} (-h + 4) = 4.$$

Since both $f(2 + 0)\ f = f(2 - 0)$, therefore $\lim_{x\to 2} f(x)$ exists and is equal to 4.

Example 13:

Evaluate the limits: $\lim_{x\to 0} \frac{1+3x^2}{x}$.

Solution:

We have $\lim_{x\to 0} \frac{1+3x^2}{x}$

We get, R.H.L. *i.e.*, $f(0 + 0) = \lim_{h\to 0} f(0 + h) = \lim_{h\to 0} f(h)$

$$= \lim_{h\to 0} \frac{1+3h^2}{h}$$

$$= \lim_{h\to 0} \left(\frac{1}{h} + 3h\right) = \infty$$

Again, L.H.L. *i.e.*, $f(0 - 0) = \lim_{h\to 0} f(0 - h) = \lim_{h\to 0} f(-h)$

$$= \lim_{h\to 0} \frac{1+3h^2}{-h}$$

$$= \lim_{h\to 0} \left(-\frac{1}{h} - 3h\right) = -\infty$$

R.H.L. $\neq$ L.H.L

Since f(0 + 0) ≠ f(0 – 0), therefore $\lim_{x\to 0} f(x)$ does not exist.

Example 14:

Evaluate: $\lim_{x\to 1}\dfrac{x^3-1}{x^2-1}$.

Solution:

We have $\lim_{x\to 1}\dfrac{x^3-1}{x^2-1} = \lim_{x\to 1}\dfrac{(x-1)(x^2+x+1)}{(x-1)(x+1)}$

$$\lim_{x\to 1}\frac{x^2+x+1}{x+1} = \frac{3}{2} \text{ Ans.}$$

Example 15:

Evaluate: $\lim_{x\to 2}\dfrac{x^2+3x+2}{x-2}$

Solution:

We have, R.H.L. *i.e.*, $f(2 + 0) = \lim_{h\to 0} f(2 + h)$

$$= \lim_{h\to 0}\frac{(2+h)^2+3(2+h)+2}{2+h-2} = \lim_{h\to 0}\frac{12+7h+h^2}{h}$$

$$= \lim_{h\to 0}\left(\frac{12}{h}+7+h\right) = \infty$$

Also, L.H.L. *i.e.*, $f(2 - 0) = \lim_{h\to 0} f(2 - h)$

$$= \lim_{h\to 0}\frac{(2-h)^2+3(2-h)+2}{2-h-2} = \lim_{h\to 0}\frac{12-7h+h^2}{-h}$$

$$= \lim_{h\to 0}\left(-\frac{12}{h}+7-h\right) = -\infty$$

Since f(2 + 0) ≠ f(2 – 0), therefore $\lim_{x\to 2} f(x)$ does not exist.

Example 16:

Evaluate: $\lim_{x\to 0}\dfrac{(1+x)^{1/3}-(1+x)^{1/3}}{x}$

Solution:

We have $\lim_{x\to 0}\dfrac{(1+x)^{1/3}-(1-x)^{1/3}}{x} = \lim_{h\to 0}\dfrac{(1+h)^{1/3}-(1-h)^{1/3}}{h}$

$$= \lim_{h\to 0} \frac{1+\frac{1}{3}h + ... - (1-\frac{1}{3}h+...)}{h} = \lim_{h\to 0} \frac{\frac{2}{3}h + ...}{h} = \frac{2}{3}.$$

Example 17:

Evaluate: $\lim_{x\to 0} e^{-1/x}$.

Solution:

We have $\lim_{x\to 0} e^{-1/x}$ let $f(x) = e^{1/x}$.

R H L. $f(0 + 0) = \lim_{h\to 0} f(0 + h) = \lim_{h\to 0} f(h) = \lim_{h\to 0} e^{-1/h} - \infty$

$$= \lim_{h\to 0} \frac{1}{e^{1/h}} = 0.$$

Again, L.H.L. $f(0 - 0) = \lim_{h\to 0} f(-h) = \lim_{h\to 0} e^{1/h} - \infty$

Since $f(0 + 0) \neq f(0 - 0)$ therefore $\lim_{x\to 0} e^{-1/x}$ does not exist.

EXERCISES

Evaluate:

1. $\lim_{x\to 1} \frac{(x-1)\sqrt{2-x}}{x^2-1}$.
2. $\lim_{x\to 1/2} \frac{8x^3-1}{5x^2-5x+1}$.
3. $\lim_{x\to 1} \frac{x^m-1}{x^n-1}$ (m and n are integers).
4. $\lim_{x\to 0} \frac{\sqrt{1+x^2}-1}{x}$.
5. $\lim_{x\to 0} \frac{\sqrt{1+x}-1}{x^2}$.
6. $\lim_{x\to 0} \frac{\sqrt{x-b}-\sqrt{a-b}}{x^2-a^2}$ (a > b).
7. $\lim_{x\to 0} \frac{3\sqrt{1+x^2}-4\sqrt{1-2x}}{x+x^2}$.

8. $\lim\limits_{x\to 1} \dfrac{3\sqrt{7+x^3}-\sqrt{3-x^2}}{x-1}$.

9. $\lim\limits_{x\to 0} \dfrac{1\_\sin x-\cos x}{1-\sin x-\cos x}$.

10. $\lim\limits_{y\to a}\left[\sin\dfrac{y-a}{2}.\tan\dfrac{\pi 4}{2a}\right]$.

11. Examine the continuity at the origin of the function $f(x)$ defined by:

$$f(x)=\begin{cases}\dfrac{|x|}{x} & x\neq 0\\ 1 & x=a\end{cases}$$

12. Using the idea of graph, examine the continuity of the function $f(x)$ defined below at $x=2$

$$f(x)=\begin{cases}1+x & x\leq 2\\ 5-x & x>2\end{cases}$$

13. According to the demand law $p=\dfrac{a}{x}$ - c, where a and c are positive constants, show that there is some demand no matter how large the price. Is there any limit to the extent of demand for small prices? Also show that total revenue falls steadily as output increases.

14. The total cost C of producing x items is given by:

$$C=\begin{cases}1000+5x & 0\leq x\leq 500\\ 2000+4x & 500<x\leq 2000\end{cases}$$

Show that the cost function C is discontinuous at $x = 500$.

15. Show that the following function is continuous at $x = 3$

$$f(x)=\begin{cases}\dfrac{x^2-x-6}{x-3} & x\neq 3\\ 5 & x=3\end{cases}$$

16. If the function given below is continuous at $x = 1$, find the value of K.

$$f(x)=\begin{cases}\dfrac{x^2-3x+2}{x-1} & x\neq 1\\ K & x=1\end{cases}$$

17. If the function given below is continuous at x = -1, find the value of C.

$$f(x) = \begin{cases} \dfrac{x^2 - 2x - 3}{x+1} & x \neq 1 \\ \text{C} & x = 1 \end{cases}$$

18. If the function given below is continuous at $x = 2$, find the value of k.

$$f(x) = \begin{cases} 2x - 1 & x < 2 \\ k & x = 2 \\ x + 1 & x > 2 \end{cases}$$

19. Show that $y = \dfrac{x^2 - 1}{x - 1}$ is continuous except at x = 1. For what value of y at x = 1, be the function becomes continuous everywhere.

3

Minima and Maxima

FIRST AND SECOND ORDER CONDITIONS FOR EXTREME VALUES

From the above figures, it is clear that a function has a maximum at a point if its curve is concave from below and has a minimum at a point if its curve is convex from below; and consequently,

(i) a sufficient condition for maxima is $\frac{dy}{dx} = 0$ and $\frac{d^2y}{dx^2} < 0$

(ii) and for minima is $\frac{dy}{dx} = 0$ and $\frac{d^2y}{dx^2} > 0$ at the point

In Economics, the conditions $\frac{dy}{dx} = 0$ and $\frac{d^2y}{dx^2} < 0$ (or $\frac{d^2y}{dx^2} > 0$) are usually called as the *first-order condition* and *second order condition* respectively for the occurrence of extreme values. Problems such as maximum utility, maximum profits, minimum cost may be effectively solved with these conditions.

Remark 1:

The set of conditions for the function $y = f(x)$ to attain a maximum at $x = a$ are:

(i) $\frac{dy}{dx} = 0$ at $x = a$; and (ii) $\frac{d^2y}{dx^2} < 0$ at $x = a$

Similarly, the set of conditions for the function $y = f(x)$ to attain a minimum at $x = a$ are:

(i) $\frac{dy}{dx} = 0$ at $x = a$; and (ii) $\frac{d^2y}{dx^2} > 0$ at $x = a$

Remark 2:

For a function $y = f(x)$, if both $\frac{dy}{dx}$ and $\frac{d^2y}{dx^2}$ are zeros at $x = a$, then the point $x = a$ may be a point of inflexion of the curve; provided $\frac{d^2y}{dx^2} \neq 0$.

THE EXISTENCE OF A MAXIMUM OR A MINIMUM OF f(x, y) AT x = a, y = b.

From the definition it is obvious that we shall have a maximum or a minimum of f(x, y) at x = a, y = b if the expression f(a + h, b + k) – f(a, b) is of invariable sign for all sufficiently small independent values of h and k provided both of them are not equal to zero. If the sign of f(a + h, b + k) – f(a, b) is negative, we shall have a maximum of f(x, y) at x = a, y = b. If is positive, f(x, y) has a minimum at x = a, y = b.

By Taylor's Theorem for two variables, we have

$$f(a+h, b+k) = f(a, b) + \left(h\frac{\partial f}{\partial x} + k\frac{\partial f}{\partial y}\right)_{\substack{x=a\\y=b}}$$

$$+\frac{1}{2!}\left(h^2\frac{\partial^2 f}{\partial x^2} + 2hk\frac{\partial^2 f}{\partial x\,\partial y} + k^2\frac{\partial^2 f}{\partial y^2}\right)_{\substack{x=a\\y=b}} + \ldots$$

$$\therefore\ f(a+h, b+k) - f(a, b) = h\left(\frac{\partial f}{\partial x}\right)_{\substack{x=a\\y=b}} + k\left(\frac{\partial f}{\partial y}\right)_{\substack{x=a\\y=b}}$$

+ terms of the second and higher orders in h and k. ...(1)

By taking h and k, sufficiently small, the first degree terms in h and k can be made to govern the sign of the right hand side and therefore of the left hand side of (1).

Thus, the sign of [f(a + h, b + k) – f(a, b)]

$$= \text{the sign of } \left[h\left(\frac{\partial f}{\partial x}\right)_{\substack{x=a\\y=b}} + k\left(\frac{\partial f}{\partial y}\right)_{\substack{x=a\\y=b}}\right] \qquad \ldots(2)$$

Taking k = 0, we find that if $\left(\frac{\partial f}{\partial x}\right)_{\substack{x=a\\y=b}} \neq 0$, the right hand side of (2) changes sign when h changes sign. Therefore f(x, y) cannot have a maximum or minimum at x = a, y = b if $\left(\frac{\partial f}{\partial x}\right)_{\substack{x=a\\y=b}} \neq 0$. Similarly, taking h = 0, we can see that f(x, y) cannot have a maximum or a minimum at

$$x = a,\ y = b \text{ if } \left(\frac{\partial f}{\partial y}\right)_{\substack{x=a\\y=b}} \neq 0 .$$

Thus, *a set of necessary conditions that f(x, y) should have a maximum or a minimum at x = a, y = b is that*

$$\left(\frac{\partial f}{\partial x}\right)_{\substack{x=a\\y=b}} = 0 \quad \textit{and} \quad \left(\frac{\partial f}{\partial y}\right)_{\substack{x=a\\y=b}} = 0$$

The above conditions are necessary but not sufficient for the existence of maxima or minima.

FUNCTIONS OF TWO INDEPENDENT VARIABLE

Let f(x, y) be any function of two independent variables x and y supposed to be continuous for all values of these variables in the neighbourhood of their values a and b respectively. Then f(a, b) is said to be a *maximum or a minimum* value of f(x, y) according as f(a + h, b + k) is *less or greater* than f(a, b) for all sufficiently small independent values of h and k, positive or negative, provided both of them are not equal to zero.

SUFFICIENT CONDITION FOR MAXIMA OR MINIMA

Let $r = \left(\frac{\partial^2 f}{\partial x^2}\right)_{\substack{x=a\\y=b}}$, $s = \left(\frac{\partial^2 f}{\partial x \partial y}\right)_{\substack{x=a\\y=b}}$ and $t = \left(\frac{\partial^2 f}{\partial y^2}\right)_{\substack{x=a\\y=b}}$.

If $\left(\frac{\partial^2 f}{\partial x}\right)_{\substack{x=a\\y=b}} = 0$ and $\left(\frac{\partial f}{\partial y}\right)_{\substack{x=a\\y=b}} = 0$, *i.e.*, if the necessary conditions existence of maxima or minima are satisfied, we have

$$f(a + h, b + k) - f(a, b) = \frac{1}{2!}(rh^2 + 2shk + tk^2) + R_3,$$

where R_3 consists of terms of third and higher orders in h and k. For sufficiently small values of h and k, the sign of

$$\frac{1}{2}(rh^2 + 2shk + tk^2) + R_3$$

is the same as that of

$$rh^2 + 2shk + tk^2.$$

Case I: $r - s^2 > 0$: In this case obviously neither r nor t can be zero. Therefore we write

$$rh^2 + 2shk + tk^2 = \frac{1}{r}[r^2h^2 + 2srhk + rtk^2]$$

$$= \frac{1}{r}\left[(rh+sk)^2 + (rt-s^2)k^2\right].$$

Since $rt - s^2$ is positive, therefore

$$(rh + sk)^2 + (rt - s^2)k^2$$

is positive for all values of h and k except when

$rh + sk = 0$, $k = 0$ *i.e.* when $h = 0$, $k = 0$ which is obviously not possible.

Thus, in this case the expression $rh^2 + 2shk + tk^2$ will have the *same* sign for all values of h and k. This sign is determined by the sign of r.

Thus f(x, y) will have a maximum or a minimum at $x = a$, $y = b$ if $rt > s^2$. Further f(x, y) is a maximum or a minimum according as r is negative or positive.

Case II: $rt - s^2 < 0$: In this case if $r \neq 0$, we can write

$$rh^2 + 3shk + tk^2 = \frac{1}{r}\,[(rh + sk)^2 + (rt - s^2)k^2].$$

If $k = 0$, $h \neq 0$, the sign of this expression will be the same as that of r. But if $k \neq 0$, $rh + sk = 0$, the sign of this expression will be opposite to that of r since $rt - s^2$ is negative. Thus in this case the expression $rh^2 + 2shk + tk^2$ is not of invariable sign.

A similar argument can be given if $t \neq 0$.

In case $r = 0$ as well as $t = 0$, we have

$$rh^2 + 2shk + tk^2 = 2shk,$$

which obviously does not keep the same sign for all values of h and k.

Thus, f(x, y) will have neither a maximum nor a minimum at $x = a$, $y = b$, if $rt < s^2$.

Case II: $rt - s^2 = 0$. If $r \neq 0$, we can write

$$rh^2 + 2shk + tk^2 = \frac{1}{r}\,[(rh + sk)^2 + (rt - s^2)k]$$

$$= \frac{1}{r}(rh+sk)^2. \qquad [\because rt - s^2 = 0].$$

This expression becomes zero when $rh + sk = 0$. Therefore the nature of the sign of

$$f(a + h, b + k) - f(a, b)$$

depends upon the consideration of R_3. The case is, therefore, doubtful.

In case r = 0, we must have s = 0, because of the condition $rt - s^2 = 0$.

$\therefore \quad rh^2 + 2shk + tk^2 = tk^2,$

which is zero when k = 0 whatever h may be. The case is given doubtful.

Thus, if $rt - s^2 = 0$, the case is doubtful and further investigation is needed to determine whether f(x, y) is a maximum or a minimum at x = a, y = b, or not.

FUNCTIONS OF A SINGLE INDEPENDENT VARIABLE

A function f(x) is said to be maximum at x = a, if there exists a positive number δ such that

$$f(a + h) < f(a)$$

for all values of h, other than zero, in the interval $(-\delta, \delta)$.

A function f(x) is said to be minimum at x = a, if there exists a positive number δ such that

$$f(a + h) > f(a)$$

for all values of h, other than zero, in the interval $(-\delta, \delta)$.

Maximum and minimum values of a function are also called its extreme values of turning values and the points at which they are attained are called points of maxima and minima.

The points at which a function has extreme values are called turning points.

PROPERTIES OF MAXIMA AND MINIMA

(i) A function y = f(x) is maximum at x = a, if dy/dx changes sign from +ive to –ive as x passed through a.

(ii) A function y = f(x) is minimum at x = a, if dy/dx changes sign from –ive to +ive as x passes through a.

(iii) There may be several maximum or minimum values of the same function.

(iv) At least one maximum or one minimum must lie between two equal values of a function.

(v) If the sign of dy/dx does not change while x passes through, a they y is neither maximum nor minimum at x = a.

(vi) Maximum and Minimum values must *occur alternately.*

SEVERAL INDEPENDENT VARIABLE OF FUNCTIONS

Let f(x, y, z, ...) be any function of several independent variables z, y, z... supposed to be continuous for all values of these variables in the neighbourhood of their values a, b, c,... respectively. Then (a, b, c,...) is said to be a maximum or a minimum value of f(x, y, z,...) for all sufficiently small independent values of h, k, *l*..., positive or negative, provided they are not all zero.

FINDING THE MAXIMA AND MINIMA OF A FUNCTION THREE INDEPENDENT VARIABLES WORKING RULE

Suppose f(x, y, z) is a given function of three independent variables x, y and z. Find $\partial f/\partial x$, $\partial f/\partial y$ and $\partial f/\partial z$ and solve the simultaneous equations $\partial f/\partial x = 0$, $\partial f/\partial y = 0$ and $\partial f/\partial z = 0$. All the triads (a, b, c) of the values of x, y and z obtained on solving these equations will give the stationary values of f(x, y, z) *i.e.*, will give the points at which the function f(x, y, z) may be a maximum or a minimum.

To discuss the maximum or minimum of f(x, y, z) at any point (a, b, c) obtained on solving the equations $\partial f/\partial x = 0$, $\partial f/\partial y = 0$ and $\partial f/\partial z = 0$, we find the values at this point of the six partial derivatives of second order of f(x, y, z) symbolically denoted as follows:

$$A = \frac{\partial^2 f}{\partial x^2}, B = \frac{\partial^2 f}{\partial y^2}, C = \frac{\partial^2 f}{\partial z^2}, F = \frac{\partial^2 f}{\partial y \partial z}, G = \frac{\partial^2 f}{\partial z \partial x} \text{ and } H = \frac{\partial^2 f}{\partial x \partial y}.$$

If the expressions

$$A, \begin{vmatrix} A & H \\ H & B \end{vmatrix}, \begin{vmatrix} A & H & G \\ H & B & F \\ G & F & C \end{vmatrix}$$

be all positive, we shall have a minimum of f(x, y, z) at (a, b, c) and if these expressions be alternately negative and positive, we shall have a maximum of f(x, y, z) at (a, b, c), whilst if these conditions are not satisfied, we shall in general have neither a maximum nor a minimum of f(x, y, z) at (a, b, c).

MAXIMUM OR MINIMUM VALUES

Necessary Conditions: *A necessary condition for f(x) to be maximum or a minimum at x = a is that f'(a) = 0.*

Let f(x) be a given function of x and suppose f(x) can be expanded in the neighbourhood of x = a by Taylor's Theorem.

$$\text{Then } f(a + h) = f(a) = hf'(a) = \frac{h^2}{2!} f''(a) + ... + \frac{h^n}{n!} f^n(a + \theta h).$$

$$\therefore\ f(a+h) - f(a) = hf'(a) + \frac{h^2}{2!} f''(a) + ... + \frac{h^n}{n!} f^n(a + \theta h). \quad ...(1)$$

By definition, f(x) is maximum at x = a if f(a + h) – f(a) < 0 for all h

$\neq 0$ in $(-\delta, \delta)$, where δ is a sufficiently small +ive number. Similarly f(x) is minimum at x = a if $f(a + h) - f(a) > 0$ for all $h \neq 0$ in $(-\delta, \delta)$, where δ is a sufficiently small +ive number. Thus, for f(x) to be a maximum or a minimum at x = a, the expression f(a + h) – f(a) must be of invariable sign (negative in the case of maximum and positive in the case of minimum) for all sufficiently small non-zero values of h, positive or negative.

Now, when h is made sufficiently small, the sign of the right hand side of (1) and therefore the sign of f(a + h) – f(a) is ultimately governed by the sign of hf'(a) as it is the term of lowest degree in h in (1). But if $f'(a) \neq 0$, the sign of hf'(a) will change with a change of sign of h. *Therefore the necessary condition that f(x) should have a maximum or a minimum at x = a is that f'(a) = 0.*

Stationary Values

Definition: *A function f(x) is said to be stationary at x = a if f'(a) = 0.*

Thus for a function f(x) to be a maximum or a minimum at x = a at must be stationary at x = a.

Sufficient Conditions for Maximum or Minimum Values

Now, if f'(a) = 0, then (1) becomes

$$f(a+h)-f(a)=\frac{h^2}{2!}f''(a)+\frac{h^3}{3!}f'''(a)+...+\frac{h^n}{n!}f^n(a+\theta h). \qquad ...(2)$$

Now, when h is made sufficiently small, the sign of the right hand side of (2) and therefore the sign of f(a + h) – f(a) is ultimately governed by the sign of $\frac{1}{2}h^2f''(a)$ as it the term of lowest degree in h in (2). But h^2 is always +ive. Therefore $\frac{1}{2}h^2f''(a)$ and f''(a) are of the same sign. Hence, if $f'(a) \neq 0$, then for all sufficiently small non-zero values of h the expression f(a + h) – f(a) is of invariable sign and its sign is the same as that of f''(a).

Therefore, there is a **maximum** *of f(x) at x = a if f'(a) = 0 and f''(a)* **is negative.**

Similarly, there is a **minimum** *of f(x) at x = a if f'(a) = 0 and f''(a)* **spositive.**

Note: If f''(a) is also equal to zero, then we can show that for a maximum or a minimum of f(x) at x = a, we must have f'''(a) = 0. Again, if $f^{iv}(a)$ is negative, there will be a maximum at x = a and if $f^{iv}(a)$ is positive, there will be a minimum at x = a.

In general if $f'(a) = f''(a) = f'''(a) = ... = f^{(n-1)}(a) = 0$ and $f^{(n)}(a) \neq 0$, then n must be an even integer for maximum or minimum. Also for a maximum $f^{(n)}(a)$ must be negative and for a minimum $f^n(a)$ must be positive.

APPLICATION OF MAXIMA AND MINIMA TO GEOMETRICAL AND OTHER PROBLEMS

Sometimes the principle of maxima and minima is applied to geometrical and other problems. In such cases we first construct from the given conditions of the problem the function whose maximum and minimum values are to be determined. If this function contains more than one variable, it is just possible to express it in terms of a single variable by eliminating the other variables with the help of the conditions of the problem. Now we apply the rules for finding maxima or minima to this resulting function of a single variable.

MAXIMA AND MINIMA OF f(x) WORKING RULE

1. Find f'(x) and equate it to zero.
2. Solve the resulting equation for x. Let its roots be $a_1, a_2, ...$ Then f(x) is stationary at $x = a_1, a_2, ...$ Thus, $x = a_1, a_2, ...$ are the only points at which f(x) can be a maximum or a minimum.
3. Find f''(x) and substitute in it by turns $x = a_1, a_2, ...$
4. If $f''(a_1)$ is –ive we have a maximum at $x = a_1$. If $f''(a_1)$ is +ive, we have a minimum at $x = a_1$.
5. If $f''(a_1) = 0$, find f'''(x) and put $x = a_1$ in it. If $f'''(a_1) \neq 0$, there is neither a maximum nor a minimum at $x = a_1$. If $f'''(a_1) = 0$, find $f^{iv}(x)$ and put $x = a_1$ in it. If $f^{iv}(a_1)$ is – ive, we have a maximum at $x = a_1$; if it is positive, there is a minimum at $x = a_1$. If $f^{iv}(a_1)$ is zero, we must find $f^{v}(x)$, and so on.

   Repeat the above process for each root of the equation f'(x) = 0.

SOLVED EXAMPLES

Example 1:

Divide a number a into three parts such that their product will be maximum.

Solution:

Let u = xyz, ...(1)

where the variables x, y and z are connected by the relation

$$x + y + z = a. \qquad ...(2)$$

From (1), log u + log x + log y + log z.

$$\therefore \quad \left(\frac{1}{u}\right)du = \left(\frac{1}{x}\right)dx + \left(\frac{1}{y}\right)dy + \left(\frac{1}{z}\right)dz.$$

For a maximum or a minimum of u, we have du = 0

$$\Rightarrow \qquad \left(\frac{1}{x}\right)dx + \left(\frac{1}{y}\right)dy + \left(\frac{1}{z}\right)dz. \qquad ...(3)$$

Also differentiating the equation (2), we have

$$dx + dy + dz = 0. \qquad ...(4)$$

Multiplying (3) by 1 and (4) by λ, and adding and then equating the coefficients of dx, dy, dz to zero, we get

$$\frac{1}{x} + \lambda = 0, \frac{1}{y} + \lambda = 0, \frac{1}{z} + \lambda = 0.$$

From these, we get $x = y = z = -\frac{1}{\lambda}$.

Putting these values of x, y, z in (2), we get

$$-\frac{3}{\lambda} = a \text{ or } \lambda = -\frac{3}{a}.$$

$\therefore$ u is stationary at the point given by

$$x = y = z = \frac{a}{3}.$$

Now regard x and y as independent variables and z as a function of x and y given by (2).

From (1), $\log u = \log x + \log y + \log z$.

$$\therefore \qquad \frac{1}{u}\frac{\partial u}{\partial x} = \frac{1}{x} + \frac{1}{z}\frac{\partial z}{\partial x}$$

But from eqn. (2) $1, + \left(\frac{\partial z}{\partial x}\right) = 0$

$$\text{or} \qquad \frac{\partial z}{\partial x} = -1$$

$$\frac{1}{u}\frac{\partial u}{\partial x} = \frac{1}{x} - \frac{1}{z}$$

so tnat $\frac{1}{u}\frac{\partial^2 u}{\partial x^2} - \frac{1}{u^2}\left(\frac{\partial u}{\partial x}\right)^2 = -\frac{1}{x^2} + \frac{1}{z^2}\frac{\partial z}{\partial x} = -\frac{1}{x^2} - \frac{1}{z^2}.$

But at the stationary point, we have $\frac{\partial u}{\partial x} = 0$.

$\therefore$ at the stationary point $\left(\frac{a}{3}, \frac{a}{3}, \frac{a}{3}\right)$, we have

$$\frac{\partial^2 u}{\partial x^2} = -u\left[\frac{1}{x^2} + \frac{1}{z^2}\right] = -xyz\left[\frac{1}{x^2} + \frac{1}{z^2}\right]$$

which is negative $x = y = z = \frac{a}{3}$.

Hence u is maximum when $x = y = z = \frac{a}{3}$ and the maximum value of $u = \left(\frac{a}{3}\right)^3$.

Example 2:

In a plane triangle ABC, find the maximum value of u = cos A cos B cos C.

Solution:

We have $u = \cos A \cos B \cos C$, ...(1)

where the variables A, B and C are connected by the relation

$$A + B + C = \pi. \quad ...(2)$$

From eqn. (1), $\log u = \log \cos A + \log \cos B = \log \cos C$.

$$\therefore \qquad \frac{1}{u} du = -\tan A\, dA - \tan B\, dB - \tan C\, dC.$$

For a maximum or a minimum of u, we must have du = 0

$$\Rightarrow \qquad \tan A\, dA + \tan B\, dB + \tan C\, dC = 0. \quad ...(3)$$

Also differentiating (2), we get

$$dA + dB + dC = 0. \quad ...(4)$$

Multiplying eqns. (3) by 1 and (4) by λ, and adding and then equating to zero the coefficients of dA, dB and dC, we get

$$\tan A + \lambda = 0, \tan B + \lambda = 0, \tan C + \lambda = 0$$

$$\text{or} \qquad -\lambda = \tan A = \tan B = \tan C$$

$$\text{or} \qquad A = B = C = \frac{\pi}{3}, \text{ from (2).}$$

Thus u is stationary when $A = B = C = \frac{\pi}{3}$ *i.e.*, the triangle is equilateral.

Now to show that the stationary value of u given by

$$A = B = C = \frac{\pi}{3} \text{ is maximum.}$$

Let us regard A and B is independent variables and C as a function of A and B given by (2).

From (1), $\log u = \log \cos A + \log \cos B + \log \cos C$.

$$\therefore \qquad \frac{1}{u}\frac{\partial u}{\partial A} = -\tan A - \tan C \cdot \frac{\partial C}{\partial A}.$$

Also differentiating (2) partially w.r.t. A taking B as constant, we get

$$1 + \left(\frac{\partial C}{\partial A}\right) = 0 \text{ or } \frac{\partial C}{\partial A} = -1.$$

$$\therefore \qquad \frac{1}{u}\frac{\partial u}{\partial A} = -\tan A + \tan C,$$

so that $\dfrac{1}{u}\dfrac{\partial^2 u}{\partial^2 A} - \dfrac{1}{u^2}\left(\dfrac{\partial u}{\partial A}\right)^2 = -\sec^2 A + \sec^2 C \cdot \dfrac{\partial C}{\partial A}$

$$= -(\sec^2 A + \sec^2 C).$$

But at the stationary point $\dfrac{\partial u}{\partial A} = 0.$

$\therefore$ at the stationary point found above, we have

$$\frac{\partial^2 u}{\partial A^2} = -u\left(\sec^2 A + \sec^2 C\right)$$

$$= -\cos A \cos B \cos C\,(\sec^2 A + \sec^2 A),$$

which is –ive for $A = B = C = \dfrac{\pi}{3}.$

Hence u is maximum when $A = B = C = \dfrac{\pi}{3}$ and the maximum value of

$$u = \left(\cos\frac{\pi}{3}\right)^3 = \left(\frac{1}{2}\right)^3 = \frac{1}{8}.$$

Example 3(a):

Show that the maximum and minimum of the radii vectors of the sections of the surface

$$(x^2 + y^2 + z^2) = \frac{x^2}{a^2} + \frac{y^2}{b^2} + \frac{z^2}{c^2}$$

by the plane $\lambda x + \mu y + \gamma z = 0$

are given by the equation

$$\frac{a^2\lambda^2}{1-a^2r^2} + \frac{b^2\mu^2}{1-b^2r^2} + \frac{c^2v^2}{1-c^2r^2} = 0.$$

Solution:

We have to find the maximum and minimum values of r, where

$$r^2 = x^2 + y^2 + z^2. \qquad ...(1)$$

Also the variables x, y, z are connected by the relations

$$\frac{x^2}{a^2} + \frac{y^2}{b^2} + \frac{z^2}{c^2} = \left(x^2 + y^2 + z^2\right)^2 = r^4 \qquad ...(2)$$

and $\quad \lambda x + \mu y + vz = 0. \qquad ...(3)$

From (1), $2r\,dr = 2x\,dx + 2y\,dy + 2z\,dz.$

For a maximum or a minimum of r, we have

$$dr = 0 \Rightarrow x\,dx + y\,dy + z\,dz = 0. \qquad ...(4)$$

Differentiating (2), we get

$$\frac{2x}{a^2}dx + \frac{2y}{b^2}dy + \frac{2z}{c^2}dz = 4r^3dr.$$

But for a maximum or a minimum of r, we have dr = 0.

$$\therefore \qquad \frac{x}{a^2}dx + \frac{y}{b^2}dy + \frac{z}{c^2}dz = 0. \qquad ...(5)$$

Also differentiating (3), we get

$$\lambda dx + \mu dy + v dz = 0. \qquad ...(6)$$

Multiplying (4) by 1, (5) by λ_1 and (6) by λ_2, and adding and then equating to zero the coefficients of dx, dy, dz, we get

$$x + \frac{x}{a^2}\lambda_1 + \lambda\lambda_2 = 0 \qquad ...(7)$$

$$y + \frac{y}{b^2}\lambda_1 + \mu\lambda_2 = 0 \qquad ...(8)$$

$$z + \frac{z}{c^2}\lambda_1 + v\lambda_2 = 0. \qquad ...(9)$$

$\therefore$ from eqn. (7), we have $x - \frac{a}{a^2} \cdot \frac{1}{r^2} + \lambda\lambda_2 = 0$

or $$x = \frac{a^2r^2\lambda\lambda_2}{1 - a^2r^2}.$$

Similarly from (8) and (9), we have

$$y = \frac{b^2r^2\mu\lambda_2}{1 - b^2r^2} \text{ and } z = \frac{c^2r^2v\lambda_2}{1 - c^2r^2}.$$

Substituting these values of x, y, z in $\lambda x + \mu y + vz = 0$, we get

$$\frac{a^2r^2\lambda^2\lambda_2}{1 - a^2r^2} + \frac{b^2r^2\mu^2\lambda_2}{1 - b^2r^2} + \frac{c^2r^2v^2\lambda_2}{1 - c^2r^2} = 0$$

or $$\frac{a^2\lambda^2}{1 - a^2r^2} + \frac{b^2\mu^2}{1 - b^2r^2} + \frac{c^2v^2}{1 - c^2r^2} = 0. \qquad ...(10)$$

The equation (10) gives the maximum and minimum values of r.

Example 3(b):

Discuss the maxima and minima of the function

u = sin x sin y sin z,

where x, y, z are the angles of a triangle.

Solution:

We have u = sin x sin y sin z, ...(1)

where x + y + z = π. ...(2)

For a maximum or a minimum of u, we must have

$$du = \cos x \sin y \sin z \, dx + \sin x \cos y \sin z \, dy + \sin x \sin y \cos z \, dz = 0 \quad ...(3)$$

Also form (2), we have

$$dx + dy + dz = 0. \quad ...(4)$$

Multiplying (3) by 1 and (4) by λ and adding and then equating to zero the coefficients of dx, dy, dz, we get

$$\cos x \sin y \sin z + \lambda = 0,$$

$$\sin x \cos y \sin z + \lambda = 0,$$

and $\sin x \sin y \cos z + \lambda = 0.$

From these, we get

$$-\lambda = \cos x \sin y \sin z = \sin x \cos y \sin z = \sin x \sin y \cos z$$

or $\cot x = \cot y = \cot z$ (dividing by sin x sin y sin z)

i.e., $x = y = z = \frac{\pi}{3}$, from (2).

Thus, u is stationary when $x = y = z = \frac{\pi}{3}$.

Let us now find the nature of this stationary value of u.

Since variables x, y and z are connected by the relation (2), only two of them may be regarded as independent.

Let us regard x and y as independent and z to be dependent on them by the relation (2).

Then from (1),

$$\frac{\partial u}{\partial x}\frac{\partial u}{\partial x} = \sin y \sin z \cos x + \sin x \sin y \cos z \frac{\partial z}{\partial x}.$$

Also form (2),

$$1 + \frac{\partial z}{\partial x} = 0 \text{ or } \frac{\partial z}{\partial x} = -1.$$

∴ $\frac{\partial u}{\partial x} = \sin y \sin z \cos x - \sin x \sin y \cos z$

and $\frac{\partial^2 u}{\partial x^2} = -\sin y \sin z \sin x - \sin y \cos x \cos z \frac{\partial z}{\partial x}$

$$- \cos x \sin y \cos z + \sin x \sin y \sin z \frac{\partial z}{\partial x}$$

$$= -2 \sin x \sin y \sin z - 2 \sin y \cos x \cos z.$$

Also $\dfrac{\partial^2 u}{\partial x\, \partial y} = \cos y \sin z \cos x + \sin y \cos x \cos z \dfrac{\partial z}{\partial y}$

$$- \sin x \cos y \cos z + \sin x \sin y \sin z \frac{\partial z}{\partial y}$$

$$= \cos y \sin z \cos x - \sin y \cos x \cos z$$
$$- \sin x \cos y \cos z + \sin x \sin y \sin z.$$

Hence putting $x = y = z = \dfrac{\pi}{3}$, we get

$$r = \frac{\partial^2 u}{\partial x^2} = -2 \cdot \frac{\sqrt{3}}{2} \cdot \frac{\sqrt{3}}{2} \cdot \frac{\sqrt{3}}{2} - 2 \cdot \frac{\sqrt{3}}{2} \cdot \frac{1}{2} \cdot \frac{1}{2} = -\frac{3\sqrt{3}}{4} - \frac{\sqrt{3}}{4} = -\sqrt{3},$$

$$s = \frac{\partial^2 u}{\partial x\, \partial y} = \frac{\sqrt{3}}{8} - \frac{\sqrt{3}}{8} - \frac{\sqrt{3}}{8} - \frac{3\sqrt{3}}{8} = -\frac{\sqrt{3}}{2}.$$

Also by symmetry, $t = \dfrac{\partial^2 u}{\partial y^2} = -\sqrt{3}$.

$\therefore$ at the stationary point $x = y = z = \dfrac{\pi}{3}$, we have

$$rt - s^2 = 3 - \left(\frac{3}{4}\right) = \frac{9}{4} \text{ which is } > 0$$

and $\quad r = -\sqrt{3}$ which is < 0.

$\therefore$ at the stationary point $x = y = z = \dfrac{\pi}{3}$, u is maximum.

Hence u is maximum when $x = y = z = \dfrac{\pi}{3}$ and the maximum value of

$$u = \left(\frac{\sqrt{3}}{2}\right)^3 = \frac{3\sqrt{3}}{8}.$$

Example 4:

Find the points where

$$u = ax^p + by^q + cz^r$$

has extreme values subject to the condition

$$x^l + y^m + z^n = k.$$

Solution:

We have $u = ax^p + by^q + cz^r$, ...(1)

where the variables x, y, z are connected by the relation

$$x^l + y^m + z^n = k. \qquad ...(2)$$

For a maximum or a minimum of u, we have

$$du = 0 \Rightarrow apx^{p-1}\,dx + bqy^{q-1}\,dy + crz^{r-1}\,dz = 0. \qquad ..(3)$$

Also differentiating (2), we get

$$lx^{l-1}\,dx + my^{m-1}\,dy + nz^{n-1}\,dz = 0. \qquad ...(4)$$

Multiplying (3) by 1 and (4) by λ, and adding and then equating to zero the coefficients of dx, dy, dz, we get

$$apx^{p-1} + \lambda l x^{l-1} = 0, \qquad ...(5)$$

$$bqy^{q-1} + \lambda m y^{m-1} = 0, \qquad ...(6)$$

and $$cr\,z^{r-1} + \lambda n\,z^{n-1} = 0. \qquad ...(7)$$

From (5), we have

$$ap\,x^{p-1} = -\lambda l\,x^{l-1} \text{ or } ap\,x^{p-l} = -\lambda l$$

or $$\frac{x^{p-l}}{l/pa} = -\lambda.$$

Similarly from (6) and (7), we have

$$\frac{y^{q-m}}{\frac{m}{qb}} = -\lambda \text{ and } \frac{z^{r-n}}{\frac{n}{rc}} = -\lambda.$$

Hence, the values of x, y, z for which u has extreme values are given by

$$\frac{x^{p-l}}{\frac{l}{pa}} = \frac{y^{q-m}}{\frac{m}{qb}} = \frac{z^{r-m}}{\frac{n}{rc}}.$$

Example 5(a):

Prove that of all rectangular parallelopipeds of the same volume the cube has the least surface.

Solution:

Let x, y, z be the dimensions of the rectangular parallelopiped, V be its volume and S be its surface. Then

$$S = 2xy + 2yz + 2zx \qquad ...(1)$$

and xyz = V = some constant. ...(2)

For a maximum or minimum of S, we have

$$dS = 2(y + z)\,dx + 2(z + x)\,dy + 2(x + y)\,dz = 0$$

i.e., $$(y + z)\,dx + (z + x)\,dy + (x + y)\,dz = 0 \qquad ...(3)$$

Also differentiating (2) and observing that V is constant, we have

$$yz\,dx + zx\,dy + xy\,dz = 0. \qquad ...(4)$$

Multiplying (3) by 1 and (4) by λ, and adding and then equating to zero the coefficients of dx, dy and dz, we get

$$(y + z) + \lambda yz = 0, \quad ...(5)$$

$$(z + x) + \lambda zx + 0, \quad ...(6)$$

and $$(x + y) + \lambda xy = 0.. \quad ...(7)$$

These give $-\lambda = \frac{1}{y} + \frac{1}{z} = \frac{1}{z} + \frac{1}{x} = \frac{1}{x} + \frac{1}{y}$.

$\therefore \quad \frac{1}{y} - \frac{1}{x} = 0$ or $x = y$.

Similarly $y = z$.

Hence for a stationary value of S, we have

$$x = y = z = V^{1/3}, \text{ from (2).}$$

Thus S is stationary when the rectangular parallelopiped is a cube.

Let us now find the nature of this stationary value of S.

Here S is a function of three variables x, y, z which are connected by the relation (2) so that only two variables are independent. Let us regard x and y as independent variables and z to be dependent on them.

Then from (1), $\frac{\partial S}{\partial x} = 2y + 2y\frac{\partial z}{\partial x} + 2z + 2x\frac{\partial z}{\partial x}$.

Also from (2), $yz + xy\frac{\partial z}{\partial x} = 0$ *i.e.*, $\frac{\partial z}{\partial x} = -\frac{z}{x}$.

$$\therefore \quad \frac{\partial S}{\partial x} = 2y = \frac{2yz}{x} + 2z - 2z = 2y - \frac{2yz}{x}$$

and $\frac{\partial^2 S}{\partial x^2} = \frac{2yz}{x^2} - \frac{2y}{x} \cdot \frac{\partial z}{\partial x} = \frac{2yz}{x^2} + \frac{2yz}{x^2} = \frac{4yz}{x^2} = 4$ at $x = y = z$.

Similarly by symmetry $\frac{\partial^2 S}{\partial y^2} = 4$ at $x = y = z$.

Also $$\frac{\partial^2 S}{\partial x \partial y} = 2 - \frac{2z}{x} - \frac{2y}{x}\frac{\partial z}{\partial y}.$$

But differentiating (2) partially w.r.t. y taking x as constant, we get

$$xz + xy\frac{\partial z}{\partial y} = 0 \text{ or } \frac{\partial z}{\partial y} = -\frac{z}{y}.$$

$$\therefore \quad \frac{\partial^2 S}{\partial x \partial y} = 2 - \frac{2z}{x} - \frac{2y}{x}\left(-\frac{z}{y}\right) = 2 - \frac{2z}{x} + \frac{2z}{x} = 2.$$

That at the stationary point $x = y = V^{1/3}$, we have

$$r = \frac{\partial^2 S}{\partial x^2} = 4,\ s = \frac{\partial^2 S}{\partial x\,\partial y} = 2 \text{ and } t = \frac{\partial^2 S}{\partial y^2} = 4.$$

$\therefore$ $rt - s^2 = 4 \times 4 - 2^2 = 12$ which is > 0.

Also $r = 4$ which is > 0.

Hence the stationary value of S given by $x = y = z = V^{1/3}$ is a minimum.

Thus of all rectangular parallelopipeds of the same volume, the cube has the least surface.

Example 5(b):

Find the maximum or minimum value of $x^2 + y^2 + z^2$*, subject to the conditions*

$$lx + my + nz = 1,\ l'x + m'y\ n'z = 1.$$

Solution:

Let $u = x^2 + y^2 + z^2$, ...(1)

where the variables x, y and z connected by the relations

$lx + my + nz = 1$, ...(2)

and $l'x + m'y + n'z = 1$. ...(3)

For a maximum or a minimum of u, we have du = 0

$\Rightarrow$ $2xdx + 2ydy + 2\,zdz = 0$

$\Rightarrow$ $xdx + ydy + zdz = 0$. ...(4)

Also differentiating (2) and (3), we get

$ldx + n'dy + ndz = 0$ ...(5)

and $l'dx + m'dy + n'dz = 0$. ...(6)

Multiplying (4) by 1, (5) by λ and (6) μ, and adding and then equating to zero the coefficients of dx, dy and dz, we get

$x + l\lambda + l'\mu = 0$, ...(7)

$y + m\lambda + m'\mu = 0$, ...(8)

and $z + n\lambda + n'\mu = 0$. ...(9)

Multiplying the equations (7), (8) and (9) by x, y and z respectively and adding, we get

$u + \lambda.z + \mu.1 = 0$. ...(10)

Again multiplying the equations (7), (8) and (9) by l, m and n respectively and adding, we get

$1 + \lambda\Sigma l^2 + \mu\ \Sigma l l'^2 = 0$. ...(11)

Next multiplying the equations (7), (8) and (9) by l', m' and n' respectively and adding, we get

$1 + \lambda\ \Sigma l l' + \mu\ \Sigma l'^2 = 0$. ...(12)

Now eliminating λ and μ from (10), (11) and (12), we get

$$\begin{vmatrix} u & 1 & 1 \\ 1 & \Sigma l^2 & \Sigma ll' \\ 1 & \Sigma ll' & \Sigma l'^2 \end{vmatrix} = 0.$$

The above equation gives the maximum or minimum value of u.

Note: If we wish to find an explicit expression for the extreme value of u and also wish to say whether it is maximum or minimum we proceed as follows:

Solving the equations (11) and (12) for λ and μ, we get

$$\frac{1}{\Sigma ll' - \Sigma l'^2} = \frac{\mu}{\Sigma ll' - \Sigma l^2} = \frac{1}{\Sigma l^2 \Sigma l'^2 - (\Sigma ll')^2}$$

But when a, b, c are all +ive, r is –ive and so the stationary value is a true maximum in this case. Also when a, b, c are all –ive, r is +ive and so the stationary value is a true minimum in this case.

or $$\lambda = \frac{\Sigma ll' - \Sigma l'^2}{\Sigma (mn' - m'n)^2} \text{ and } \mu = \frac{\Sigma ll' - \Sigma l'^2}{\Sigma (mn' - m'n)^2}$$

[$\because \Sigma l^2 \Sigma l'^2 - (\Sigma ll')^2 = \Sigma (mn' - m'n)^2$, by Lagrange's identity]

Putting these values of λ and μ in (10), the maximum or minimum value of u is given by

$$u = -\lambda - \mu = \frac{\Sigma l^2 + \Sigma l'^2 - 2\Sigma ll'}{\Sigma (mn' - m'n)^2} = \frac{\Sigma (l - l')^2}{\Sigma (mn' - m'n)^2}.$$

Example 5(c):

Show that the maximum and minimum values of $u = ax^2 + by^2 + cz^2 + 2fyz + 2gzx + 2hxy$ subject to the conditions $lx + my + nz = 0$ and $x^2 + y^2 z^2 = 1$ are given by the equation

$$\begin{vmatrix} a-u, & h & g & l \\ h & b-u & f & m \\ g & f & c-u & n \\ l & m & n & 0 \end{vmatrix} = 0.$$

Solution:

We have $u = ax^2 + by^2 + cz^2 + 2fyz + 2gzx + 2hxy$, ...(1)

where the variables x, y and z are connected by the relations

$$lx + my + nz = 0 \quad ...(2)$$

and $$x^2 + y^2 + z^2 = 1. \quad ...(3)$$

For a maximum or a minimum of u, we have du = 0

$\Rightarrow 2(ax + gz + hy) + 2(by + fz + hx)dy + 2(cz + fy + gx)dz = 0$

$\Rightarrow (ax + hy + gz)dx + (hx + by + fz)dy + (gx + fy + cz)\, dz = 0.$ ...(4)

Also differentiating (2) and (3), we get

$l\, dx + m\, dy + n\, dz = 0$ ...(5)

and $x\, dx + y\, dy + z\, dz = 0.$ ...(6)

Multiplying (4), by 1, (5) by λ and (6) by μ, and adding and then equating to zero the coefficient of dx, dy and dz, we get

$(ax + hy + gz)\, \lambda l + \mu x = 0$ ...(7)

$(hx + by + fz)\, \lambda m + \mu y = 0$ (8)

and $(gx + fy + cz)\, \lambda n + \mu z = 0$ ...(9)

Multiplying eqns. (7), (8) and (9) by x, y and z respectively and adding, we get

$u + \lambda.0 + \mu.1 = 0$ or $m = -u.$

Putting $\mu = -u$ in (7), (8) and (9), we get

$(a - u\, x + hy + gz + \lambda l = 0,$ ....(10)

$hx + (b - y) + fz + \lambda m = 0,$ ...(11)

and $gx + fy + (c - u)\, z + \ln = 0.$ ...(12)

Also the relation (2) can be written as

$lx + my + nz + \lambda.0 = 0.$ ...(13)

Eliminating x, y, z and λ from (10), (11), (12) and (13), we get

$$\begin{vmatrix} a-u, & h & g & 1 \\ h & b-u & f & m \\ g & f & c-u & n \\ 1 & m & n & 0 \end{vmatrix} = 0,$$

which gives the maximum and minimum values of u.

To Find the Nature of This Stationary Valve of u

Since there are two relations amongst the variables x, y and z, therefore only one variable will be independent. Let it be x. Then

$$\frac{du}{dx} = \frac{\partial x}{\partial x} \cdot \frac{dy}{dx} + \frac{\partial u}{\partial z} \cdot \frac{dz}{dx}$$

$$= 2x + 2y \cdot \frac{dy}{dx} + 2z \cdot \frac{dz}{dx}.$$

Now differentiating (2) and (3) w.r.t. x, we get

$$l + m\frac{dy}{dx} + n\frac{dz}{dx} = 0$$

and $\qquad l'+m'\dfrac{dy}{dx}+n'\dfrac{dz}{dx}=0.$

Solving these, we get

$$\frac{\frac{dy}{dx}}{nl'-n'l}=\frac{\frac{dz}{dx}}{m'l-l'm}=\frac{1}{mn'-m'n}.$$

$$\therefore \quad \frac{dy}{dx}=\frac{nl'-n'l}{mn'-m'n} \text{ and } \frac{dz}{dx}=\frac{m'l-l'm}{mn'-m'n}.$$

$$\therefore \quad \frac{du}{dx}=2x+2y\cdot\frac{nl'-n'l}{mn'-m'n}+2z\cdot\frac{m'l-l'm}{mn'-m'n}$$

and so $\quad \dfrac{d^2u}{dx^2}=2+2\dfrac{dy}{dx}\cdot\dfrac{nl'-n'l}{mn'-m'n}+2\dfrac{dz}{dx}\cdot\dfrac{m'l-l'm}{mn'-m'n}$

$$=2+2\left(\frac{nl'-n'l}{mn'-m'n}\right)^2+2\left(\frac{m'l-l'm}{mn'-m'n}\right)^2,$$

which is +ive.

Hence, the stationary value of u found above is the minimum value.

Example 6:

Find the maximum and minimum values of y where $y = (x-1)(x-2)^2$.

Solution:

Let $f(x) = y = (x-1)(x-2)^2$.

Then

$$f'(x)=\frac{dy}{dx}=(x-2)^2+2(x-1)(x-2)$$
$$=(x-2)(3x-4).$$

For a maximum or minimum of y, $f'(x)=\dfrac{dy}{dx}$

i.e., $(x-2)(3x-4)=0$, whence $x = 2, \dfrac{4}{3}$.

Also $f''(x)=\dfrac{d^2y}{dx^2}=6x-10.$

Now $f''(2) = 2$, *i.e.*, +ive $\Rightarrow$ f(x) has a minimum at x = 2.

Again $f''\left(\dfrac{4}{3}\right)=-2$, *i.e.*, –ive

$\Rightarrow$ f(x) has a maximum at $x=\dfrac{4}{3}$.

Example 7:

Prove that the maximum and minimum values of the function $y = (ax^2 + 2bx + c)/(Ax^2 + 2Bx + C)$ are those values of y for which $(ax^2 + 2bx + c) - (Ax^2 + 2Bx + C)$ is perfect square.

Solution:

We have $y = \dfrac{(ax^2 + bx + c)}{(Ax^2 + 2Bx + C)}$.

$\therefore \quad y(Ax^2 + 2Bx + C) = ax^2 + 2bx + c.$ ...(1)

Differentiating (1) w.r.t. x, we get

$$\left(\frac{dy}{dx}\right)(Ax^2 + 2Bx + C) + y(2Ax + 2B) = 2ax + 2b. \quad ...(2)$$

For a maximum or a minimum of y we have $\dfrac{dy}{dx} = 0$. Putting $\dfrac{dy}{dx} = 0$ in (2), we get

$$y(Ax + B) = ax + b. \quad ...(3)$$

Thus, the maximum and the minimum values of y are obtained by eliminating x between (1) and (3). From (3), we get

$x = -\dfrac{(By - b)}{(Ay - a)}$. Also (1) can be written as

$$x^2(Ay - b) + 2x(By - b) + Cy - c = 0. \quad ...(4)$$

Putting $x = -\dfrac{(By - b)}{(Ay - a)}$ in (4), we get

$$\frac{(By - b)^2}{Ay - a} - \frac{2(By - b)^2}{Ay - a} + Cy - c = 0$$

or $\quad (By - b)^2 - (Cy - c)(Ay - a) = 0,$ ...(5)

giving the maximum and minimum values of y.

Now the condition that the expression

$$(ax^2 + 2bx + c) - y(Ax^2 + 2Bx + C)$$

or $\quad x^2(a - Ay) + 2x(b - By) + c - Cy$

be a perfect square is that

$$4(b - By)^2 - 4(a - Ay)(c - Cy) = 0$$

or $\quad (By - b)^2 - (Ay - a)(Cy - c).$ ...(6)

Since the equations (5) and (6) are identical, therefore maximum and the minimum values of y are those for which

$$(ax^2 + 2bx + c) - y\,(Ax^2 + 2Bx + C)$$

is a perfect square.

Example 8(a):

Prove that $x^2 \log (1/x)$ has a maximum value when $x = e^{-1/2}$.

Solution:

Let $\quad y = x^2 \log\left(\dfrac{1}{x}\right) = x^2 (\log 1 - \log x) = -x^2 \log x.$

Then $\quad \dfrac{dy}{dx} = -2x\log x - x^2\left(\dfrac{1}{x}\right)$

$$= -x(2\log x + 1) \qquad ...(1)$$

For a maximum or a minimum of y, we must have

$$\frac{dy}{dx} = 0 \text{ i.e., } -x(2\log x + 1) = 0$$

i.e., $\quad x = 0$, or $2\log x + 1 = 0$ giving $\log x = -\dfrac{1}{2}$ or $x = e^{-1/2}$.

Now differentiating (1) again, we have

$$\frac{d^2y}{dx^2} = -(2\log x + 1) - x.\left(\frac{2}{x}\right) = -(2\log x + 1) - 2.$$

When $x = e^{-1/2}$, we have $2\log x + 1 = 0$.

Therefore when $x = e^{-1/2}$,

we have $\dfrac{d^2y}{dx^2} = -2$, which is negative.

Hence y is maximum when $x = e^{-1/2}$.

Note that when $x = 0$, the value of y does not exist.

Example 8(b):

If $dy/dx = (x - a)^{2n} (x - b)^{2p+1}$, where n and p are positive integers, show that $x = a$ gives neither a maximum nor a minimum value of y, but $x = b$ gives a minimum.

Solution:

For a maximum or a minimum of y, we must have

$$\frac{dy}{dx} = 0$$

i.e., $\quad (x - a)^{2n} (x - b)^{2p+1} = 0$

i.e., $\quad x = 0$

or $\quad x = b.$

To test nature of y at x = b: When x is slightly less than b, dy/dx is negative as 2p + 1 is odd. Also when x is slightly greater than b, dy/dx is positive. Thus, dy/dx changes sign from negative to positive while x passes through the value b. Hence y is minimum at x = b.

To test the nature of y at x = a: Here we see that the sign of dy/dx when x is slightly less than a is the same as when x is slightly greater than a since 2n is even. Thus dy/dx does not change sign while x passes through the value a. Therefore y is neither a maximum nor a minimum at x = a.

Example 12:

Find the maximum and minimum values of $\sin^2\theta + \sin^2\phi$, *where* $\theta + \phi = \alpha$.

Solution:

Let $u = \sin^2\theta + \sin^2\phi$, where $\theta + \phi = \alpha$.

Then $u = \sin^2\theta + \sin^2(\alpha - \theta)$.

$$\therefore \frac{du}{d\theta} = 2\sin\theta\cos\theta - 2\sin(\alpha-\theta)\cos(\alpha-\theta)$$

$$= \sin 2\theta - \sin(2\alpha - 2\theta).$$

For a maximum or a minimum of u, we have $\frac{du}{d\theta} = 0$

i.e., $\sin 2\theta - \sin(2\alpha - 2\theta) = 0$,

i.e., $\sin 2\theta = \sin(2\alpha - 2\theta)$

i.e., $2\theta = n\pi + (-1)^n(2\alpha - 2\theta)$.

If n is odd, we have $2\theta = n\pi - 2\alpha + 2\theta$ which does not give any value of θ. If n is even, say 2m, we have $2\theta = 2m\pi + 2\alpha - 2\theta$ which gives $\theta = (m\pi + \alpha)/2$. Thus u is stationary when $\theta = (m\pi + \alpha)/2$. Also then $\phi = \alpha - \theta = \alpha - (m\pi + \alpha)/2 = (\alpha - m\pi)/2$.

Now $d^2u/d\theta^2 = 2\cos 2\theta + 2\cos(2\alpha - 2\theta) = 4\cos\alpha\cos(2\theta - \alpha)$. When $\theta = (m\pi + \alpha)/2$, we have $d^2u/d\theta^2 = 4\cos\alpha\cos m\pi$.

If m is odd, say 2r + 1, then $d^2u/d\theta^2 = -4\cos\alpha$ which is –ive. Therefore u is maximum when $\theta = \{(2r+1)\pi + \alpha\}/2$.

If m is even, say 2r, then $d^2u/d\theta^2 = 4\cos\alpha$ which is +ive. Therefore u is minimum when $\theta = (2r\pi + \alpha)/2$.

The maximum value of $u = \sin^2\frac{1}{2}(m\pi + \alpha) + \sin^2\frac{1}{2}(\alpha - m\pi)$, (where m is odd) $= \cos^2\frac{1}{2}\alpha + \cos^2\frac{1}{2}\alpha = 2\cos^2\frac{1}{2}\alpha = 1 + \cos\alpha$.

The minimum value of $u = \sin^2(m\pi + \alpha)/2 + \sin^2(\alpha - m\pi)/2$, (where m is even) $= \sin^2\alpha/2 + \sin^2\alpha/2 = 2\sin^2\alpha/2 = 1 - \cos\alpha$.

Example 8(c):

The sum of the sides of a rectangle is constant. If the area is to be a maximum show that the rectangle must be a square.

Solution:

Let x and y be the sides of the rectangle.

Then $x + y = c$, a constant. ...(1)

Now area $A = xy - x(c - x) = cx - x^2$.

We have $\frac{dA}{dx} = c - 2x$.

For a maximum or minimum of A, $\frac{dA}{dx} = c - 2x$ *i.e.*, $c - 2x = 0$

i.e., $x = \frac{c}{2}$.

Now $\frac{d^2A}{dx^2} = -2$, which is negative therefore A is maximum when $x = \frac{c}{2}$. When $x = \frac{c}{2}$, we have from (1), $y = x - \frac{1}{2}c = \frac{c}{2}$. Thus, for the area of the rectangle to be maximum, we have $x = y = \frac{c}{2}$ *i.e.*, the rectangle must be a square.

Example 9:

The sum of one number and three times a second number is 60. Among the possible numbers satisfying the condition, find the pair whose product is maximum.

Solution:

Let the two numbers be x and y. Then as given,

$x + 3y = 60$. ...(1)

Now let $z = xy$ = the product of the two numbers.

Then $z = x.\frac{1}{3}(60 - x)$, from (1)

$= 20x - \frac{1}{3}x^2$.

We have $\frac{dz}{dx} = 20 - \frac{2}{3}x$.

For a maximum or a minimum of z, $\frac{dz}{dx} = 0$

i.e., $20 - \frac{2}{3}x = 0$

i.e., $x = 30$.

Putting $x = 30$ in eqn. (1), we get $3y = 60 - 30$

i.e., y = 10.

Now $\frac{d^2z}{dx^2} = -\frac{2}{3}$ which is –ive.

Hence z is maximum when x = 30, y = 10.

Example 10:

A given quantity of metal is to be cast into a half cylinder i.e., with a rectangular base and semi-circular ends. Show that in order that the total surface area may be a minimum, the ratio of the length of the cylinder to the diameter of the ends is $\pi/(\pi + 2)$.

Solution:

Let l be the length and r be the radius of each end of the half cylinder.

If V is the volume of the half cylinder, then $V = \frac{1}{2}\pi r^2 l$ = constant, so that $l = 2V/(\pi r^2)$.

Now let S be the total surface of the half cylinder. Then S = area of the rectangular base + curved surface + area of semi-circular ends

$$= 2rl + \pi rl + 2\left(\frac{1}{2}\pi r^2\right) = (2 + \pi)\, rl + \pi r^2$$

i.e., $$S = (2 + p)\left(\frac{2V}{\pi r}\right) + \pi r^2, \text{ from (1).}$$

For a maximum or a minimum of S, $\frac{dS}{dr} = 0$.

Now $$\frac{dS}{dr} \equiv \frac{-2(2+\pi)V}{\pi r^2} + 2\pi r = 0 \text{ gives}$$

$$-\frac{2(2+\pi)\frac{1}{2}\pi r^2 l}{\pi r^2} + 2\pi r = 0, \qquad \left[\because V = \frac{1}{2}\pi r^2 l\right]$$

or $$\left(\frac{l}{2r}\right) = \frac{\pi}{(\pi+2)}.$$

Also $$\frac{d^2S}{dr^2} = \frac{\pi+2}{\pi} \cdot \frac{4V}{r^3} + 2\pi = +\text{ive}.$$

Hence the total surface S is minimum when

$$\left(\frac{l}{2r}\right) = \frac{\pi}{(\pi+2)}$$

i.e., when the ratio of the length of the cylinder to the diameter of the ends is $\pi : \pi + 2$.

Example 11(a):

The strength of a rectangular beam varies as the product of its breadth and the square of its depth. Find the dimensions of the strongest beat that can be cut from a round log of diameter 2a.

Solution:

Let x be the breadth and y the depth of the beam, then as given

$$x^2 + y^2 = 4a^2. \qquad ...(1)$$

If S denotes the strength of the beam,

then $S \propto xy^2$, or $S = kxy^2$,

where k is some constant.

Now $\quad S = kx(4a^2 - x^2)$, from (1),

$$\therefore \quad \frac{dS}{dx} = k\left(4a^2 - 3x^2\right).$$

For a maximum or a minimum of S, $\frac{dS}{dx} = 0$.

i.e. $\quad 4a^2 - 3x^2 = 0$

i.e., $\quad x = \frac{2a}{\sqrt{3}}$ because x cannot be –ive.

Now $\frac{d^2S}{dx^2} = -6kx$ = negative, when $x = \frac{2a}{\sqrt{3}}$. Therefore S is maximum

when $\quad x = \frac{2a}{\sqrt{3}}$. When $x = \frac{2a}{\sqrt{3}}$, we have from eqn. (1),

$$y = \sqrt{\left(4a^2 - \frac{1}{3} \cdot 4a^2\right)} = 2a\sqrt{\left(\frac{2}{3}\right)}.$$

Hence for the strongest be a, depth = $\sqrt{\left(\frac{2}{3}\right)}$ · diameter of the log and

breadth = $\left(\frac{1}{\sqrt{3}}\right)$ · diameter of the log.

Example 11(b):

Assuming that the strength of a beam of rectangular cross section varies as the product of its breadth and the cube of its depth, find the breadth of the strongest beam which can be cut from a circular log of diameter a.

Solution:

Let x be the breadth and y be the depth of the beam.

Then $x^2 + y^2 = a^2$. ...(1)

Let S = the strength of the beam = kxy^3, where k is some constant.

Now $S = ky^3 (a^2 - y^2)^{1/2}$, [from (1)].

Since S is positive, therefore S is maximum or minimum according as S^2 is maximum or minimum.

Let $\quad u = S^2 = k^2y^6(a^2 - y^2) = k^2a^2y^6 - k^2y^8.$

We have $\dfrac{du}{dy} = 6k^2a^2y^5 - 8k^2y^7 = k^2y^5\,(6a^2 - 8y^2).$

For a maximum or a minimum of u, $\dfrac{du}{dy} = 0$

i.e., $k^2y^5\,(6a^2 - 8y^2) = 0$

i.e., $6a^2 - 8y^2 = 0$ since y cannot be zero,

i.e., $y^2 = \dfrac{6a^2}{8} = \dfrac{3a^2}{4}.$

Now $\dfrac{d^2u}{dy^2} = 30k^2a^2y^4 - 56k^2y^6 = 2k^2y^4\,(15a^2 - 28y^2) = -$ ive, when $y^2 = 3a^2/4$. Therefore u is maximum, and hence S is maximum, when $y^2 = 3a^2/4$. When $y^2 = 3a^2/4$, we have from (1), $x^2 = a^2/4$ or $x = a/2$. Hence for the strongest beam, the breadth $= a/2$.

Example 11(c):

Show that the height of the cylinder of maximum volume that can be inscribed in a sphere of radius a is $2a/\sqrt{3}$.

Solution:

Let a be the radius of the sphere and O be its centre. Then OA = a where A is the one of the lower point of contact of the cylinder and the sphere. Also let MN and AN represent the height and radius of the cylinder.

Let $MN = x$ then $ON = \dfrac{x}{2};$

$\therefore\ AN^2 = OA^2 - ON^2 = a^2 - \dfrac{x^2}{4}.$

Thus V = volume of the cylinder

$$= \pi AN^2.MN = \pi\left(a^2 - \frac{x^2}{4}\right)x = \pi\left(a^2x - \frac{x^3}{4}\right).$$

Differentiating, we get $\dfrac{dV}{dx} = \pi\left(a^2 - \dfrac{3x^2}{4}\right).$

For max. or min. of V, $\dfrac{dV}{dx} = 0$

or $\qquad \left(a^2 - \dfrac{3x^2}{4}\right) = 0$

or $\quad 3x^2 = 4a^2$

or $\quad x = \dfrac{2a}{\sqrt{3}}.$

Now $d^2V/dx^2 = -3\pi x/2 = -$ive when $x = 2a/\sqrt{3}$. Hence volume of the cylinder is max. when height of the cylinder is $2a/\sqrt{3}$.

Example 11(d):

Prove that the minimum radius vector of the curve $a^2/x^2 + b^2/y^2 = 1$ is of length $a + b$.

Solution:

Let r be the radius vector of the curve. Then changing the equation of the curve to polar coordinates by putting $x = r\cos\theta$ and $y = r\sin\theta$, we get

$$\frac{a^2}{(r^2\cos^2\theta)} + \frac{b^2}{(r^2\sin^2\theta)} = 1$$

or $\quad r^2 = a^2\sec^2\theta + b^2\operatorname{cosec}^2\theta = z$, say.

Now, $|r|$ is max. or min. according as r^2 *i.e.*, z is maximum or minimum. We have $dz/d\theta = 2a^2\sec^2\theta\tan\theta - 2b^2\operatorname{cosec}^2\theta\cot\theta$.

For a maximum or a minimum of z, we have $dz/d\theta = 0$

i.e., $\quad 2a^2\sec^2\theta\tan\theta - 2b^2\operatorname{cosec}^2\theta\cot\theta = 0$

i.e., $\quad \dfrac{(\sec^2\theta\tan\theta)}{(\operatorname{cosec}^2\theta\cot\theta)} = \dfrac{b^2}{a^2}$

i.e., $\quad \tan^4\theta = \dfrac{b^2}{a^2}$

i.e., $\quad \tan^2\theta = \dfrac{b}{a}$ because $\tan^2\theta$ cannot be –ive.

Now $\dfrac{d^2z}{d\theta^2} = 4a^2\sec^2\theta\tan^2\theta + 2a^2\sec^4\theta + 4b^2\operatorname{cosec}^2\theta\cot^2\theta$

$+ 2b^2\operatorname{cosec}^4\theta = +$ive when $\tan^2\theta = \dfrac{b}{a}$.

Therefore, z or r is minimum when $\tan^2\theta = \dfrac{b}{a}$.

Now $r = \sqrt{(a^2\sec^2\theta + b^2\operatorname{cosec}^2\theta)}$

$= \sqrt{\{a^2(1+\tan^2\theta) + b^2(1+\cot^2\theta)\}}.$

Therefore, the minimum value of $r = \sqrt{\left\{a^2\left(1+\dfrac{b}{a}\right) + b^2\left(1+\dfrac{a}{b}\right)\right\}}$

$$= \sqrt{(a^2 + 2ab + b^2)} = \sqrt{(a+b)^2} = a + b.$$

Example 12:

An open rectangular tank, with a square base and vertical sides, is to be constructed of sheet metal to hold a given quantity of water. Show that the cost of the material will be least when the depth is half the width.

Solution:

Let x be the width of the tank and y be its depth. If V be the given capacity of the tank, then

$$V = x^2y$$

or $$y = \frac{V}{x^2} \qquad ...(1)$$

Now, if S be the area of the sheet required to construct the tank, then S = area of the base + area of the four vertical walls

$$= x^2 + 4xy = x^2 + 4x\left(\frac{V}{x^2}\right), \qquad \text{from (1)}$$

$$= x^2 + \frac{4V}{x}.$$

For a maximum or minimum of S, we have

$$\frac{dS}{dx} \equiv 2x - \frac{4V}{x^2} = 0$$

i.e., $2V = x^3$.

Also $$\frac{d^2S}{dx^2} = 2 + \frac{8V}{x^3} = +\text{ive}$$

for $x^3 = 2V$.

Hence S is least when $2V = x^3$

or when $2x^2y = x^3$, $\qquad [\because V = x^2y, \text{ from (1)}]$

or when $y = \frac{x^3}{(2x^2)} = \frac{x}{2}$

or when depth $= \frac{1}{2}$ width.

Example 13:

Show that if x = a is an approximate position for a maximum or minimum of f(x), then

$$f(a) = \frac{\{f'(a)\}^2}{f''(a)}$$

is, in general, a better approximation than f(a) to the maximum or minimum value.

Solution:

We have $f(x) = f[a + (x - a)]$

$$= f(a) + (x - a) f'(a) + \{(x - a)^2/2!\} f''(a) + \dots \quad \dots(1)$$

[By Taylor's Theorem]

Differentiating (1) w.r.t. x, we get

$$f'(x) = f'(a) + (x - a) f''(a) + \dots \quad \dots(2)$$

For a maximum or a minimum of f(x), we have $f'(x) = 0$.

Retaining only the first two terms on the right side of (2), we see that $f'(x) = 0$ is the same equation approximately as

$$f'(a) + (x - a) f''(a) = 0.$$

This gives $x - a = - f'(a)/f''(a)$.

Substituting this value of $x - a$ in (1), we see that

$f(a) - \{f'(a)^2/f''(a)$ is a better approximation than f(a) for the maximum or minimum value of f(x).

Example 14(a):

Find the maximum value of

$$(ax + by + cz)\, e^{-}(\alpha^2x^2 + \beta^2y^2 + \gamma^2z^2).$$

Solution:

Let $u = (ax + by + cz)\, e^{-}(\alpha^2x^2 + \beta^2y^2 + \gamma^2z^2)$.

Then $\log u = \log (ax + by + cz) - (\alpha^2x^2 + \beta^2y^2 + \gamma^2z^2)$.

$$\therefore \quad \frac{1}{u}\frac{\partial u}{\partial x} = \frac{a}{ax+by+cz} - 2\alpha^2 x, \quad \frac{1}{u}\frac{\partial u}{\partial y} = \frac{b}{ax+by+cz} - 2\beta^2 y,$$

$$\frac{1}{u}\frac{\partial u}{\partial z} = \frac{c}{ax+by+cz} - 2\gamma^2 z.$$

For a maximum or a minimum of u, we must have

$$\frac{\partial u}{\partial x} = 0, \quad \frac{\partial u}{\partial y} = 0, \quad \frac{\partial u}{\partial z} = 0.$$

These give $x(ax + by + cz) = \dfrac{a}{2\alpha^2}$ ...(1)

$$y(ax + by + cz) = \frac{b}{2\beta^2} \quad \dots(2)$$

and $$z(ax + by + cz) = \frac{c}{2\gamma^2}. \quad \dots(3)$$

Multiplying (1), (2), (3) by a, b, c respectively and adding, we get

$$(ax + by + cz)^2 = \frac{1}{2}\left(\frac{a^2}{\alpha^2}+\frac{b^2}{\beta^2}+\frac{c^2}{\gamma^2}\right),$$

so that $(ax + by + cz) = \sqrt{\left\{\frac{1}{2}\left(\frac{a^2}{\alpha^2}+\frac{b^2}{\beta^2}+\frac{c^2}{\gamma^2}\right)\right\}} = R$, say.

Then $x=\frac{a}{2\alpha^2 R},\ y=\frac{b}{2\beta^2 R},\ z=\frac{c}{2\gamma^2 R}.$

$\therefore$ u is stationary when $x=\frac{a}{2\alpha^2 R},\ y=\frac{b}{2\beta^2 R},\ z=\frac{c}{2\gamma^2 R}.$

Again $\frac{1}{u}\frac{\partial^2 u}{\partial x^2}-\frac{1}{u^2}\left(\frac{\partial u}{\partial x}\right)^2=-\frac{a^2}{(ax+by+cz)^2}-2\alpha^2.$

Now at a stationary point,

we have $\frac{\partial u}{\partial x}=0.$

$\therefore$ at the stationary point found above, we have

$\frac{\partial^2 u}{\partial x^2}=-u\left[\frac{a^2}{(ax+by+cz)^2}+2\alpha^2\right]$, which is –ive because u is positive for the values of x, y, z found above.

$\therefore$ u is maximum at the stationary point found above.

Also putting the values of x, y, z found above in the value of u, the maximum value of u

$$= R.e^{-\frac{1}{4R^2}\left(\frac{a^2}{\alpha^2}+\frac{b^2}{\beta^2}+\frac{c^2}{\gamma^2}\right)} = R.e^{-\frac{1}{4R^2}.2R^2}$$

$$= R.e^{-1/2}=\frac{R}{\sqrt{e}}=\frac{1}{\sqrt{e}}\cdot\sqrt{\left\{\frac{1}{2}\left(\frac{a^2}{\alpha^2}+\frac{b^2}{\beta^2}+\frac{c^2}{\gamma^2}\right)\right\}}$$

$$=\sqrt{\left\{\frac{1}{2e}\left(\frac{a^2}{\alpha^2}+\frac{b^2}{\beta^2}+\frac{c^2}{\gamma^2}\right)\right\}}.$$

Example 14(b):

Discuss the maximum or minimum values of u where

$$u = x^2 + y^2 + z^2 + x - 2z - xy.$$

Solution:

For a maximum or a minimum of u, we must have

$$\frac{\partial u}{\partial x}=2x-y+1=0,$$

$$\frac{\partial u}{\partial y}=-x+2y=0,$$

and $$\frac{\partial u}{\partial z}=2z-2=0.$$

These equations give $x=-\frac{2}{3}$, $y=-\frac{1}{3}$, z = 1.

$\therefore \left(-\frac{2}{3},-\frac{1}{3},1\right)$ is the only point at which u is stationary *i.e.*, at which u may have a maximum or a minimum.

Now $$\frac{\partial^2 u}{\partial x^2}=2,\ \frac{\partial^2 u}{\partial y^2}=2,\ \frac{\partial^2 u}{\partial z^2}=2,$$

$$\frac{\partial^2 u}{\partial y\,\partial z}=0, \qquad \frac{\partial^2 u}{\partial z\,\partial x}=0$$

and $$\frac{\partial^2 u}{\partial x\,\partial y}=-1.$$

If A, B, C, F, G and G denote the respective values of these six partial derivatives of second order at the point $\left(-\frac{2}{3},-\frac{1}{3},1\right)$, then

$$A = 2,\ B = 2,\ C = 2,\ F = 0,\ G = 0,\ H = -1.$$

Now we have

$$A = 2,\ \begin{vmatrix} A & H \\ H & B \end{vmatrix}=\begin{vmatrix} 2 & -1 \\ -1 & 2 \end{vmatrix}=3$$

and $$\begin{vmatrix} A & H & G \\ H & B & F \\ G & F & C \end{vmatrix}=\begin{vmatrix} 2 & -1 & 0 \\ -1 & 2 & 0 \\ 0 & 0 & 2 \end{vmatrix}=6.$$

Since these three expressions are all positive, we have a minimum of u when $x=-\frac{2}{3}$, $y=-\frac{1}{3}$, z = 1.

Example 15:

Find the maximum or minimum values of u where

$$u = axy^2z^3 - x^2y^2z^3 - xy^3z^3 - xy^2z^4.$$

Solution:

We have $\frac{\partial u}{\partial x} = ay^2z^3 - 2xy^2z^3 - y^3z^3 - y^2z^4$

$= y^2z^3 (a - 2x - y - z),$

$\frac{\partial u}{\partial y} = 2a\ xyz^3 - 2x^2yz^3 - 3xy^2z^3 - 2xyz^4$

$= xyz^3 (2a - 2x - 3y - 2z)$

and $\frac{\partial u}{\partial z} = 2axy^2z^2 - 3x^2y^2z^2 - 3xy^3z^3 - 4xy^2z^3$

$= xy^2z^2(3a - 3x - 3y - 4z).$

For a maximum or a minimum of u, we must have

$$\frac{\partial u}{\partial x} = 0,\ \frac{\partial u}{\partial y} = 0,\ \frac{\partial u}{\partial z} = 0.$$

Now one solution of the equations $\frac{\partial u}{\partial x} = 0$, $\frac{\partial u}{\partial y} = 0$, $\frac{\partial u}{\partial z} = 0$ is given by the equations

$$2x + y + z = a,\ 2x + 3y + 2z = 2a,\ 3x + 3y + 4z = 3a.$$

Solving these equations, we get $x = \frac{a}{7}$, $y = \frac{2a}{7}$, $z = \frac{3a}{7}$.

$\therefore$ u is stationary at the point $\left(\frac{a}{7}, \frac{2a}{7}, \frac{3a}{7}\right)$.

Now $A = \frac{\partial^2 u}{\partial x^2} = y^2z^3.(-2).$

At the stationary point $\left(\frac{a}{7}, \frac{2a}{7}, \frac{3a}{7}\right)$, the value of A is –ive.

$\therefore$ u has a maximum value at the point

$$\left(\frac{a}{7}, \frac{2a}{7}, \frac{3a}{7}\right).$$

Putting $x = \frac{a}{7}$, $y = \frac{2a}{7}$, $z = \frac{3a}{7}$ in the value of u, the maximum value of u at the point $\left(\frac{a}{7}, \frac{2a}{7}, \frac{3a}{7}\right)$.

Example 16:

Find the maximum value of u where

$$u = \frac{xyz}{(a+x)(x+y)(y+z)(z+b)}.$$

Solution:

We have

$$\log u = \log x + \log y + \log z - \log (a + x) - \log (x + y) - \log (y + z) - \log (z + b).$$

$$\therefore \quad \frac{1}{u}\frac{\partial u}{\partial x} = \frac{1}{x} - \frac{1}{a+x} - \frac{1}{x+y} = \frac{ay - x^2}{x(a+x)(x+y)}$$

$$\frac{\partial u}{\partial x} = \frac{(ay - x^2)u}{x(a+x)(x+y)}.$$

Similarly $\dfrac{\partial u}{\partial x} = \dfrac{(xz - y^2)u}{y(x+y)(y+z)}$

and $\dfrac{\partial u}{\partial z} = \dfrac{(by - z^2)u}{z(y+z)(z+b)}.$

Now for a maximum or a minimum of u, we must have

$$\frac{\partial u}{\partial x} = 0 \text{ i.e., } ay - x^2 = 0$$

$$\frac{\partial u}{\partial y} = 0 \text{ i.e., } xz - y^2 = 0$$

and $\dfrac{\partial u}{\partial z} = 0$ *i.e.*, $by - z^2 = 0.$

From the above equations, it follows that a, x, y, z and b are in geometrical progression. Let r be the common ratio of this geometrical progression. Then

$$ar^4 = b \text{ or } r = \left(\frac{b}{a}\right)^{1/4}.$$

Also $x = ar$, $y = ar^2$, $z = ar^3$.

Substituting these values, we get

$$u = \frac{ar.ar^2.ar^3}{a(1+r).ar(1+r).ar^2(1+r).ar^3(1+r)}$$

$$= \frac{1}{a(1+r)^4} = \frac{1}{a\left[1 + \left(\frac{b}{a}\right)^{1/4}\right]^4} = \frac{1}{\left(a^{1/4} + b^{1/4}\right)^4}$$

This gives a stationary value of u. To decide whether this value of u is a maximum or a minimum we proceed to find the second order partial derivatives of u.

We have

$$\frac{\partial^2 u}{\partial x^2}=\frac{-2xu}{x(a+x)(x+y)}+\left(ay-x^2\right)\frac{\partial}{\partial x}\left[\frac{u}{x(a+x)(x+y)}\right].$$

$\therefore$ when $x = ar$, $y = ar^2$, $z = ar^3$, we have

$$A=\frac{\partial^2 u}{\partial x^2}=\frac{-2.ar.u}{ar.a(1+r).ar(1+r)}=\frac{-2u}{a^2r(1+r)^2},$$

which is negative.

Hence the above stationary value of u is a maximum.

Example 17(a):

Discuss the maximum or minimum values of

$$u = x^3y^2(1 - x - y).$$

Solution:

We have $\frac{\partial u}{\partial x} = 3x^2y^2(1 - x - y) + x^3y^2(-1)$

$= 3x^2y^2 - 4x^3y^2 - 3x^2y^3$,

and $\frac{\partial u}{\partial y} = 2x^3y(1 - x - y) + x^3y^2(-1) = 2x^3y - 2x^4y - 3x^3y^2$.

For a maximum or minimum of u, we must have $\frac{\partial u}{\partial x} = 0$ and $\frac{\partial u}{\partial y} = 0$

i.e.,

$$x^2y^2(3 - 4x - 3y) = 0$$

and $x^3y(2 - 2x - 3y) = 0;$

$$x = 0, -\infty < y < \infty;$$

and $y = 0, -\infty < x < \infty.$

Now $r=\frac{\partial^2 u}{\partial x^2} = 6xy^2 - 12x^2y^2 - 6xy^3 = 6xy^2(1 - 2x - y),$

$$s=\frac{\partial^2 u}{\partial x\,\partial y} = 6x^2y - 8x^3y - 9x^2y^2 = x^2y(6 - 8x - 9y),$$

and $t=\frac{\partial^2 u}{\partial y^2} = 2x^3 - 2x^4 - 6x^3y = 2x^3(1 - x - 3y).$

When $x = \dfrac{1}{2}, y = \dfrac{1}{3}$, we have

$$r = 6 \cdot \frac{1}{2} \cdot \frac{1}{9}\left(1 - 2 \cdot \frac{1}{2} - \frac{1}{3}\right) = -\frac{1}{9},$$

$$s = \frac{1}{4} \cdot \frac{1}{3}\left(6 - 8 \cdot \frac{1}{2} - 9 \cdot \frac{1}{3}\right) = -\frac{1}{12}$$

and $$t = 2 \cdot \frac{1}{8}\left(1 - \frac{1}{2} - 3 \cdot \frac{1}{3}\right) = -\frac{1}{12}$$

$\therefore$ $rt - s^2 = \left(-\dfrac{1}{9}\right)\left(-\dfrac{1}{8}\right) - \left(\dfrac{1}{12}\right)^2 = \dfrac{1}{12} - \dfrac{1}{144} = +\text{ive}$. Also is –ive. Hence is maximum at $x = \dfrac{1}{2}, y = \dfrac{1}{3}$.

When $x = 0$ and $-\infty < y < \infty$, we have $rt - s^2 = 0$. Also when $y = 0$ and $-\infty < x < \infty$, we have $rt - s^2 = 0$. So these cases are doubtful and need further investigation.

Example 17(b):

Find the extreme values of xy (a – x – y).

Solution:

Let $u = xy(a - x - y)$.

Then $$\frac{\partial u}{\partial x} = ay - 2xy - y^2$$

and $$\frac{\partial u}{\partial y} = ax - x^2 - 2xy.$$

Also $$r = \frac{\partial^2 u}{\partial x^2} = -2y,\ s = \frac{\partial^2 u}{\partial y\, \partial x} = a - 2x - 2y,$$

and $$t = \frac{\partial^2 u}{\partial y^2} = -2x.$$

For a maximum or minimum of u, we have

$$\frac{\partial u}{\partial x} = 0 \text{ and } \frac{\partial u}{\partial y} = 0.$$

Thus, we have

$$\left.\begin{array}{l} ay - 2xy - y^2 = 0 \\ ax - x^2 - 2xy = 0 \end{array}\right\}.$$

These equations can be written as

$$y(a - 2x - y) = 0, \ x(a - x - 2y) = 0,$$

so that we have to consider the four pairs of equations *viz.*,

$$y = 0, x = 0, a - 2x - y = 0, x = 0;$$
$$y = 0, a - x - 2y = 0;$$
$$a - 2x - y = 0, a - x - 2y = 0.$$

Solving these, we get the following pairs of values of x and y which make the function stationary:

$$(0, 0), (0, a), (a, 0), \left(\frac{1}{3}a, \frac{1}{3}a\right).$$

For (0, 0),

$$r = 0, s = a, t = 0 \text{ so that } rt - s^2 \text{ is negative.}$$

Therefore, u has not an extreme value at (0, a).

Similarly, we may show that u has not an extreme value at (a, 0).

For $\left(\frac{1}{3}a, \frac{1}{3}a\right)$,

$$r = -\frac{2}{3}a, s = -\frac{1}{3}a, t = -\frac{2}{3}a \text{ so that } rt - s^2 \text{ is negative.}$$

Therefore u has an extreme value at $\left(\frac{1}{3}a, \frac{1}{3}a\right)$ and it will be a maximum or minimum according as, r, is negative or positive *i.e.*, according as, a, is positive or negative.

$\therefore$ the extreme value of u is $\left(\frac{1}{27}\right)a^3$.

Example 17(c):

Find the point within a a triangle such that the sum of the squares of its distances from the three vertices is a minimum.

Solution:

Let (x_r, y_r), r = 1, 2, 3 be the vertices of the triangle and (x, y) be any point inside the triangle.

Let $u = \sum_{r=1}^{3}\left[(x - x_r)^2 + (y - y_r)^2\right]$.

Then $\frac{\partial u}{\partial x} = \Sigma\, 2\,(x - x_r) = 2\,[(x - x_1) + (x - x_2) + (x - x_3)]$

and $\frac{\partial u}{\partial x} = \Sigma\, 2\,(y - y_r) = 2\,[(y - y_1) + (y - y_2) + (y - y_3)$.

Also $r=\frac{\partial^2 u}{\partial x^2}=6$, $s=\frac{\partial^2 u}{\partial x\, \partial y}=0$ and $t=\frac{\partial^2 u}{\partial y^2}=6$.

For a maximum or a minimum of u, we must have

$$\frac{\partial u}{\partial x}=0$$

$$\Rightarrow \quad 2[(x - x_1) + (x - x_2) + (x - x_3)] = 0$$

$$\Rightarrow \quad x=\frac{x_1+x_2+x_3}{3}$$

$\therefore \left(\frac{x_1+x_2+x_3}{3}, \frac{y_1+y_2+y_3}{3}\right)$ is the only point at which u may have a maximum or a minimum.

At this point, r = 6, s = 0, t = 6, so that $rt - s^2 = 36$ which is > 0.

$\therefore$ the stationary value of u at the point

$$\left(\frac{x_1+x_2+x_3}{3}, \frac{y_1+y_2+y_3}{3}\right)$$

is an extreme value.

Since r > 0, therefore this extreme value of u is a minimum.

$$\left(\frac{x_1+x_2+x_3}{3}, \frac{y_1+y_2+y_3}{3}\right),$$

which is the centroid of the given triangle.

Example 18:

If dy/dx = $x(x-1)^2(x-3)^3$, find the maximum and minimum value of y.

Solution:

For a maximum or a minimum of y, we must have

$$\frac{dy}{dx}=0 \textit{ i.e., } x(x-1)^2(x-3)^3=0$$

i.e., x = 0 or x = 1 or x = 3.

Nature of y at x = 0: When x is slightly less than 0, dy/dx is positive. Also when x is slightly greater than 0, dy/dx is negative. Thus, dy/dx changes sign from positive to negative while x passes through the value 0. Hence y is maximum at x = 0.

Nature of y at x = 1: When x is slightly less than 1, dy/dx is negative and when x is slightly greater than 1, dy/dx is again negative. Thus, dy/dx does not change sign while x passes through the value 1.

Therefore y is neither a maximum nor a minimum at x = 1.

Nature of y at x = 3: When x is slightly less than 3, dy/dx is negative and when x is slightly greater than 3, dy/dx is positive. Thus dy/dx changes sign from negative to positive while x passes through the value 3. Hence, y is minimum at x = 3.

Example 19:

Show that sin x(1 + cos x) is a maximum at x = p/3.

Solution:

Let $f(x) = \sin x(1 + \cos x) = \sin x + \frac{1}{2}\sin 2x$.

Then $f'(x) = \cos x + \cos 2x$.

For a maximum or a minimum value of f(x), f'(x) = 0

i.e., $\cos x + \cos 2x = 0$ *i.e.*, $2\cos^2 x + \cos x - 1 = 0$

i.e., $(2\cos x - 1)(\cos x + 1) = 0$.

$\therefore \cos x = \frac{1}{2}, -1$ *i.e.*, $x = \frac{\pi}{3}, \pi$.

Now $f''(x) = -\sin x - 2\sin 2x$.

$$\therefore f''\left(\frac{\pi}{3}\right) = -\sin\left(\frac{\pi}{3}\right) - 2\sin\left(\frac{2\pi}{3}\right) = -\text{ive}.$$

Hence f(x) is maximum at $x = \frac{\pi}{3}$.

Example 20:

Find the maximum value of $\frac{(\log x)}{x}$ *in* $0 < x < \infty$.

Solution:

$$\text{Let } f(x) = \frac{(\log x)}{x}. \qquad ...(1)$$

$$\text{Then } f'(x) = \frac{\left[x.\left(\frac{1}{x}\right) - \log x\right]}{x^2} = \frac{(1 - \log x)}{x^2}.$$

For a maximum or a minimum value of f(x), f'(x) = 0

i.e., $1 - \log x = 0$ *i.e.*, $x = e$.

$$\text{Now } f''(x) = (1 - \log x)\left(-\frac{2}{x^3}\right) - \left(\frac{1}{x^3}\right).$$

$\therefore f''(e) = -\frac{1}{e^3} = -$ ive. Therefore f(x) is a maximum at x = e.

Putting x = e in (1), the maximum value of $f(x) = \frac{(\log e)}{e} = \frac{1}{e}$.

Example 21(a):

Prove that $x^5 - 5x^4 + 5x^3 - 10$ has a maximum for x = 1, a minimum for x = 3, and for x = 0, it has neither a maximum nor a minimum.

Solution:

Let $f(x) = x^5 - 5x^4 + 5x^3 - 10.$

For a maximum or minimum of f(x) it is necessary that

$$f'(x) = 0$$

i.e., $5x^4 - 20x^3 + 15x^2 = 0$

i.e., $5x^2(x^2 - 4x + 3) = 0$

i.e., $5x^2(x - 3)(x - 1) = 0$

i.e., $x = 0, 1$ or 3.

Thus, f(x) is stationary at x = 0, 1 and 3.

Now $f''(x) = 20x^3 - 60x^2 + 30x.$

We have $f''(1) = 20 - 60 + 30 = -10$ *i.e.,* – ive.

Hence f(x) has a maximum value at x = 1.

Again $f''(3) = 540 - 540 + 90 = 90$ *i.e.,* +ive.

Hence f(x) has a minimum value at x = 3.

Further we have $f''(0) = 0$. So we find $f'''(x)$. We have

$$f'''(x) = 60x^2 - 12x + 30 \text{ which is } \neq 0 \text{ when } x = 0.$$

Hence at x = 0, f(x) has neither a maximum nor a minimum.

Example 21(b):

Find the largest and the smallest values of the function $x^3 - 18x^2 + 96x$ in the interval [0, 9].

Solution:

Let $f(x) = x^3 - 18x^2 + 96x.$ ...(1)

Then $f'(x) = 3x^2 - 36x + 96 = 3(x^2 - 12x + 32)$

$$= 3(x - 4)(x - 8).$$

For a maximum or minimum value of f(x), $f'(x) = 0$ *i.e.,* $(x - 4)(x - 8) = 0$ *i.e.,* x = 4, 8. Both these values lie in the interval [0, 9].

Now $f''(x) = 6x - 36 = 6(x - 6).$

At x = 4, f''(x) = –ive. Therefore, f(x) is a maximum at x = 4.

At x = 8, f''(x) = +ive. Therefore, f(x) is a minimum at x = 8.

Thus in the interval [0, 9], the function f(x) is maximum at x = 4 and is minimum at x = 8. We have $f(4) = 64 - 288 + 384 = 160$ = the maximum value of f(x) in [0, 9]. Also the minimum value of f(x) in [0, 9] is f(8) = 512 – 1152 + 768 = 128.

Now at the ends of the interval [0, 9] we have f(0) = 0, and f(9) = 135. The value 0 is less than the minimum value 128. Therefore 0 is the smallest value of f(x) in [0, 9]. The value f(9) *i.e.*, 135 is less than the maximum value 160. Therefore 160 is the largest value of f(x) in [0, 9].

Example 21(c):

Show that $\frac{x}{1+x\tan x}$ *is maximum when* $x = \cos x$.

Solution:

Let $f(x)=\frac{x}{1+x\tan x}$

Then $f'(x)=\frac{(1+x\tan x)-x\left(x\sec^2 x+\tan x\right)}{(1+x\tan x)^2}=\frac{1-x^2\sec^2 x}{(1+x\tan x)^2}$.

For a maximum or minimum value of f(x), f'(x) = 0

i.e., $1 - x^2\sec^2 x = 0$ *i.e.*, $x = \pm \cos x$.

$$\text{Now } f''(x) = (1 - x^2\sec^2 x)\frac{d}{dx}\frac{1}{(1+x\tan x)^2} + \frac{1}{(1+x\tan x)^2}(-2x\sec^2 x - 2x^2\sec^2 x\tan x)$$

$$=\left(1-x^2\sec^2 x\right)\frac{d}{dx}\frac{1}{(1+x\tan x)^2}-\frac{2x\sec^2 x}{(1+x\tan x)^2}\cdot(1+x\tan x)$$

$$=\left(1-x^2\sec^2 x\right)\frac{d}{dx}\frac{1}{(1+x\tan x)^2}-\frac{2x\sec^2 x}{1+x\tan x}.$$

When $x = \cos x$, $f''(x)=0-\frac{2}{\cos x(1+\sin x)}$ *i.e.*, –ive.

Therefore f(x) has a maximum value when x = cos x.

Example 21(d):

Show that $\sin^p\theta\cos^q\theta$ *attains a maximum when* $\theta=\tan^{-1}\sqrt{\left(\frac{p}{q}\right)}$.

Solution:

Let $y = \sin^p\theta\cos^q\theta$. For a maximum or minimum of y,

we have $\frac{dy}{d\theta}=0$

i.e., $p \sin^{p-1}\theta \cos^{q+1}\theta - q \sin^{p+1}\theta \cos^{q-1}\theta = 0$

i.e., $\sin^{p-1}\theta \cos^{q-1}\theta \,(p \cos^2\theta - q \sin^2\theta) = 0$

i.e., $\sin\theta = 0$ or $\cos\theta = 0$ or $\tan^2\theta = \dfrac{p}{q}$

i.e., $\theta = 0$ or $\theta = \dfrac{\pi}{2}$ or $\theta = \tan^{-1}\sqrt{\left(\dfrac{p}{q}\right)}$.

Now y = 0 at q = 0 and also $\theta = \dfrac{\pi}{2}$. When $0 < \theta < \dfrac{\pi}{2}$, y is +ive. Also $\tan^{-1}\sqrt{\left(\dfrac{p}{q}\right)}$ is the only value of θ lying between 0 and $\dfrac{\pi}{2}$ at which $\dfrac{dy}{d\theta} = 0$.

Hence y is maximum when $\theta = \tan^{-1}\sqrt{\left(\dfrac{p}{q}\right)}$. This is clear from the graph of y.

Example 22(a):

Prove that the maximum value of $(1/x)^x$ is $e^{1/e}$.

Solution:

Let $y = \left(\dfrac{1}{x}\right)^x$.

Then log y = x (log 1 − log x) = − x log x.

$$\therefore \quad \left(\frac{1}{y}\right)\left(\frac{dy}{dx}\right) = -1.\log x - x\left(\frac{1}{x}\right) = -(1 + \log x).$$

or
$$\frac{dy}{dx} = -y(1+\log x) = -\left(\frac{1}{x}\right)^x (1 + \log x). \qquad ...(1)$$

For a maximum or a minimum or y it is necessary that

$$\frac{dy}{dx} = 0 \text{ i.e., } -\left(\frac{1}{x}\right)^x (1+\log x) = 0$$

i.e., 1 + log x = 0 *i.e.,* log x = − 1 *i.e.,* $x = e^{-1} = \dfrac{1}{e}$.

Now differentiating (1) again, we have

$$\frac{d^2y}{dx^2} = -\left(\frac{dy}{dx}\right)(1+\log x) - y\left(\frac{1}{x}\right)$$

$$= -\left(\frac{dy}{dx}\right)(1+\log x) - \left(\frac{1}{x}\right)^x \cdot \left(\frac{1}{x}\right). \qquad ...(2)$$

When $x = \frac{1}{e}$, we have $\frac{dy}{dx} = 0$. Therefore when $x = \frac{1}{e}$, we have from (2)

$$\frac{d^2y}{dx^2} = 0 - (e)^{1/e}.e$$, which is negative.

Hence, is maximum when $x = \frac{1}{e}$ and putting $x = \frac{1}{e}$ in the value of y, the maximum value of y is $e^{1/e}$.

Example 22(b):

Find the maximum value of (x – 1) (x – 2) (x – 3).

Solution:

Let $f(x) = (x - 1)(x - 2)(x - 3) = x^3 - 6x^2 + 11x - 6.$

Then $f'(x) = 3x^2 - 12x + 11.$

For a maximum or minimum value of f(x), f'(x) = 0

i.e., $3x^2 - 12x + 11 = 0$

$$i.e.,\ x = \frac{12 \pm \sqrt{(144 - 4 \times 3 \times 11)}}{6} = \frac{6 \pm \sqrt{(36 - 33)}}{3} = 2 \pm \frac{1}{\sqrt{3}}.$$

Also $f''(x) = 6x - 12.$

Now $f''\left[2 + \left(\frac{1}{\sqrt{3}}\right)\right]$ = +ive. Therefore f(x) has minimum at $x = 2 + \left(\frac{1}{\sqrt{3}}\right)$.

Again $f''\left[2 - \left(\frac{1}{\sqrt{3}}\right)\right]$ = –ive. Therefore f(x) has a minimum at $x = 2 - \left(\frac{1}{\sqrt{3}}\right)$.

The maximum value of f(x) is $= f\left[2 - \left(\frac{1}{\sqrt{3}}\right)\right]$

$$= \left[1 - \left(\frac{1}{\sqrt{3}}\right)\right]\left(-\frac{1}{\sqrt{3}}\right)\left[-1 - \left(\frac{1}{\sqrt{3}}\right)\right]$$

$$= \left(1 - \frac{1}{3}\right)\left(\frac{1}{\sqrt{3}}\right) = \frac{2}{(3\sqrt{3})}.$$

Example 23:

Show that the point such that the sum of the squares of its distances from n given points shall be minimum, is the centre of the mean position of the given points.

Solution:

Let the n given points be (a_1, b_1, c_1), (a_2, b_2, c_2), ..., (a_n, b_n, c_n) and let (x, y, z) be the coordinates of the required point.

If u denotes the sum of the squares of the distances of (x, y, z) from the n given points, then

$$u = \Sigma[(x - a_1)^2 + (y - b_1)^2 + (z - c_1)^2]$$
$$= \Sigma(x - a_1)^2 + S(y - b_1)^2 + S(z - c_1)^2.$$

For a maximum or a minimum of u, we must have

$$\frac{\partial u}{\partial x} = 2\Sigma(x - a_1) = 2nx - 2\Sigma a_1 = 0,$$

$$\frac{\partial u}{\partial y} = 2\Sigma(y - b_1) = 2ny - 2\Sigma b_1 = 0,$$

and $$\frac{\partial u}{\partial z} = 2\Sigma(z - c_1) = 2nz - 2\Sigma c_1 = 0.$$

Solving these equations, we get

$$x = \frac{\Sigma a_1}{n},\ y = \frac{\Sigma b_1}{n},\ z = \frac{\Sigma c_1}{n}.$$

Now $$A = \frac{\partial^2 u}{\partial x^2} = 2n,\ B = \frac{\partial^2 u}{\partial y^2} = 2n,\ C = \frac{\partial^2 u}{\partial z^2} = 2n,$$

$$F = \frac{\partial^2 u}{\partial y \partial z} = 0,\ G = \frac{\partial^2 u}{\partial z \partial x} = 0,\ H = \frac{\partial^2 u}{\partial x \partial y} = 0.$$

We have A = 2n, $$\begin{vmatrix} A & H \\ H & B \end{vmatrix} = \begin{vmatrix} 2n & 0 \\ 0 & 2n \end{vmatrix} = 4n^2,$$

and $$\begin{vmatrix} A & H & G \\ H & B & F \\ G & F & C \end{vmatrix} = \begin{vmatrix} 2n & 0 & 0 \\ 0 & 2n & 0 \\ 0 & 0 & 2n \end{vmatrix} = 8n^3.$$

Since these three expressions are all positive, u is minimum when

$$x = \frac{\Sigma a_1}{n},\ y = \frac{\Sigma b_1}{n} \text{ and } z = \frac{\Sigma c_1}{n}$$

i.e., u is minimum when the point (x, y, z) is the centre of the mean position of the n given points.

Example 24:

Show that $u = (x + y + z)^3 - 3(x + y + z) - 24xyz + a^3$ *has minimum at (1, 1, 1) and maximum at (–1, –1, –1).*

Solution:

For a maximum or a minimum of u, we must have

$$\frac{\partial u}{\partial x} = 3\,(x + y + z)^2 - 3 - 24yz = 0, \quad \text{...(1)}$$

$$\frac{\partial u}{\partial y} = 3\,(x + y + z)^2 - 3 - 24zx = 0 \quad \text{...(2)}$$

and $$\frac{\partial u}{\partial z} = 3\,(x + y + z)^2 - 3 - 24xy = 0. \quad \text{...(3)}$$

Subtracting (2) from (1), we get

$$24z\,(x - y) = 0$$

which has x = z for one of its solutions.

Thus, the equations (1), (2) and (3) are satisfied when

$$x = y = z$$

Putting y = x and z = x (1), we get

$$27x^2 - 3 - 24x^2 = 0 \text{ or } 3x^2 = 3 \text{ or } x^2 = 1 \text{ or } x = \pm 1.$$

$\therefore$ x = y = z = 1 and x = y = z = – 1 are solutions of the equations (1), (2) and (3).

Hence u is stationary at the points (1, 1, 1) and (– 1, – 1, – 1).

Now $$A = \frac{\partial^2 u}{\partial x^2} = 6(x + y + z),$$

$$B = \frac{\partial^2 u}{\partial y^2} = 6(x + y + z),$$

$$B = \frac{\partial^2 u}{\partial y^2} = 6(x + y + z),$$

$$F = \frac{\partial^2 u}{\partial y\,\partial z} = 6(x + y + z) - 24x,$$

$$G = \frac{\partial^2 u}{\partial z\,\partial x} = 6(x + y + z) - 24y,$$

$$H = \frac{\partial^2 u}{\partial x\,\partial y} = 6(x + y + z) = -24z.$$

Nature of u at (1, 1, 1): At the stationary point (1, 1, 1), we have A = 18, B = 18, C = 18, F = – 6, G = – 6, H = – 6.

$\therefore$ at the point (1, 1, 1), we have

$$A = 18, \begin{vmatrix} A & H \\ H & B \end{vmatrix} = \begin{vmatrix} 18 & -6 \\ -6 & 18 \end{vmatrix} = 288$$

and
$$\begin{vmatrix} A & H & G \\ H & B & F \\ G & F & C \end{vmatrix} = \begin{vmatrix} 18 & -6 & -6 \\ -6 & 18 & -6 \\ -6 & -6 & 18 \end{vmatrix}$$

$$= 6^3.\begin{vmatrix} 3 & -1 & -1 \\ -1 & 3 & -1 \\ -1 & -1 & 3 \end{vmatrix}$$

$$= 6^3.\begin{vmatrix} 1 & 1 & 1 \\ -1 & 3 & -1 \\ -1 & -1 & 3 \end{vmatrix}, \text{ by } R_1 + R_2 + R_3$$

$$= 6^3.\begin{vmatrix} 1 & 0 & 0 \\ -1 & 4 & 0 \\ -1 & 0 & 4 \end{vmatrix}, \text{ by } R_2 - R_1 \text{ and } R_3 - R_1$$

$$= 6^3.16.$$

Since these three expressions are all positive, we have a minimum of u at the point (1, 1, 1).

Nature of u at the stationary point (–1, –1, –1).

At the stationary point (–1, –1, –1), we have

$$A = -18 = B = C,\ F = 6 = G = H.$$

∴ at the point (–1, –1, –1), we have

$$A = -18, \begin{vmatrix} A & H \\ H & B \end{vmatrix} = \begin{vmatrix} -18 & 6 \\ 6 & -18 \end{vmatrix} = 288$$

and
$$\begin{vmatrix} A & H & G \\ H & B & F \\ G & F & C \end{vmatrix} = \begin{vmatrix} -18 & 6 & 6 \\ 6 & -18 & 6 \\ 6 & 6 & -18 \end{vmatrix} = -6^3.16.$$

Since these three expressions are alternately negative and positive, we have a maximum of u at the point (–1, –1, –1).

Example 25(a):

If two variables x and y are connected by the relation $ax^2 + by^2 = ab$, *show that the maximum and minimum values of the function $u = x^2 + y^2 + xy$ will be the roots of the equation*

$$4(u - a)(u - b) = ab.$$

Solution:

We have $u = x^2 + y^2 + xy$, ...(1)

where the variables x and y are connected by the relation

$$ax^2 + by^2 = ab. \qquad ...(2)$$

For a maximum or a minimum of u, we have

$$du = 0 \Rightarrow 2x\, dx + 2y\, dy + y\, dx + x\, dy = 0$$

$$\Rightarrow \quad (2x + y)\, dx + (2y + x)\, dy = 0. \qquad ...(3)$$

Also differentiating (2), we have

$$2ax\, dx + 2by\, dy = 0 \text{ or } ax\, dx + by\, dy = 0. \qquad ...(4)$$

Multiplying (3) by 1 and (4) by λ, and adding and then equating to zero the coefficients of dx and dy, we get

$$(2x + y) + \lambda\, ax = 0 \qquad ...(5)$$

and $$(2y + x) + \lambda\, by = 0. \qquad ...(6)$$

Multiplying (5) by x and (6) by y and adding, we get

$$2(x^2 + y^2 + xy) + \lambda\,(ax^2 + by^2) = 0$$

or $$2u + \lambda\, ab = 0 \text{ or } \lambda = -\frac{2u}{ab}.$$

Putting $\lambda = -\frac{2u}{ab}$ in (5), we get

$$(2x+y) - \frac{2u}{b}x = 0$$

or $$2(b - u)\, x + by = 0. \qquad ...(7)$$

Similarly putting $\lambda = -\frac{2u}{ab}$ in (6), we get

$$(2y+x) - \frac{2u}{a}y = 0$$

or $$ax + 2\,(a - u)\, y = 0. \qquad ...(8)$$

Eliminating x and y from (7) and (8), we get

$$\begin{vmatrix} 2(b-u) & b \\ a & 2(a-u) \end{vmatrix} = 0$$

or $$4(b - u)(a - u) - ab = 0$$

or $4(u - a)(u - b) = ab$, which gives the maximum and minimum values of u.

Example 25(b):

Find the stationary values of $x^2 + y^2 + z^2$ *subject to the conditions*

$$ax^2 + by^2 + cz^2 = 1 \text{ and } lx + my + nz = 0.$$

Interpret the result geometrically.

Solution:

Let $\quad u = x^2 + y^2 + z^2,$ ...(1)

where the variables x, y and z are connected by the relations

$$ax^2 + by^2 + cz^2 = 1, \quad ..(2)$$

and $\quad lx + my + nz = 0.$

For a stationary value of u, we have

$$du = 0$$

$\Rightarrow \quad 2x\,dx + dy\,dy + 2z\,dz = 0$

$\Rightarrow \quad x\,dx + y\,dy + z\,dz = 0.$ ...(4)

Also differentiating the given relations (2) and (3), we get

$$2ax\,dx + 2by\,dy + 2cz\,dz = 0$$

i.e., $\quad ax\,dx + by\,dy + cz\,dz = 0$ ...(5)

and $\quad l dx + m\,dy + n\,dz = 0.$ ...(6)

Multiplying equations (4) by 1, (5) by λ and (6) by μ and adding, and then equating the coefficients of dx, dy, dz to zero, we get

$$x + \lambda ax + \mu l = 0 \quad ...(7)$$

$$y + \lambda by + \mu m = 0 \quad ...(8)$$

$$z + \lambda cz + \mu n = 0 \quad ...(9)$$

Multiplying the equations (7), (8) and (9) by x, y and z respectively we get

$$x^2 + y^2 + z^2 + \lambda\,(ax^2 + by^2 + cz^2) + \mu\,(lx + my + nz) = 0,$$

or $\quad u = \lambda.1 + \mu.0 = 0$, using (1), (2) and (3)

or $\quad \lambda = -u.$

Substituting for λ in the equations (7), (8) and (9), we get

$$x = \frac{\mu l}{au-1}, y = \frac{\mu m}{bu-1}, z = \frac{\mu n}{cu-1}.$$

Substitutng these values of x, y, z in (3), we get

$$\frac{\mu l^2}{au-1} + \frac{\mu m^2}{bu-1} + \frac{\mu n^2}{cu-1} = 0$$

or $$\frac{l^2}{au-1} + \frac{m^2}{bu-1} + \frac{n^2}{cu-1} = 0. \quad ...(10)$$

Hence the stationary (*i.e.*, maximum or minimum) values of u are given by the equation (10). The equation (10) is a quadratic in u and so it gives two stationary values of u.

Geometrical Interpretation

The surface $ax^2 + by^2 + cz^2 = 1$ represents an ellipsoid (or a hyperboloid) whose centre is origin, and $lx + my + nz = 0$ is a plane passing through the origin. Therefore the point (x, y, z) satisfying both the conditions (2) and (3) lies on the conic in which (2) and (3) intersect. Also $x^2 + y^2 + z^2$ gives the square of the distance of (x, y, z) from the origin which is also the centre of the conic of intersection. The maximum and minimum values of this distance are the major and minor semi-axes of the conic. So the equation (10) gives the squares of the lengths of the semi-axes of the conic of intersection.

Example 25(c):

Find the maximum and minimum values of

$$u = a^2x^2 + b^2y^2 + c^2z^2,$$

where $x^2 + y^2 + z^2 = 1$ *and* $lx + my + nz = 0.$

Solution:

We have $u = a^2x^2 + b^2y^2 + c^2z^2,$ ...(1)

where the variables x, y, z are connected by relations

$$x^2 + y^2 + z^2 = 1 \qquad ...(2)$$

and $lx + my + nz = 0$ ...(3)

For a maximum or a minimum of u, we have

$$du = 0$$

$$\Rightarrow \quad 2a^2x\,dx + 2b^2y\,dy + 2c^2z\,dz = 0$$

$$\Rightarrow \quad a^2x\,dx + b^2y\,dy + c^2z\,dz = 0. \qquad ...(4)$$

Also differentiating the two given equations (2) and (3) connecting the variables x, y and z, we get

$$2x\,dx + 2y\,dy + 2z\,dz = 0$$

i.e., $x\,dx + y\,dy + z\,dz = 0$ ...(5)

and $l dx + m\,dy + n\,dz = 0$ ...(6)

Multiplying (4), (5) and (6) by 1, λ and μ respectively and adding, and then equating to zero the coefficients of dx, dy and dz, we get

$$a^2x + \lambda x + \mu l = 0, \qquad ...(7)$$

$$b^2y + \lambda y + \mu m = 0, \qquad ...(8)$$

and $c^2z + \lambda z + \mu m = 0.$ ...(9)

Multiplying the equations (7), (8) and (9) by x, y and z respectively and adding, we get

$$u + \lambda.1 + \mu.0 = 0 \text{ or } \lambda = -u.$$

Putting $\lambda = -u$ in (7), we get

$$a^2x - ux + \mu l = 0$$

or $$x = \frac{\mu l}{u - a^2}.$$

Similarly, from (8) and (9), we get

$$y = \frac{\mu m}{u - b^2} \text{ and } z = \frac{\mu m}{u - c^2}.$$

Substituting these values of x, y, z in $lx + my + nz = 0$, we get

$$\frac{ul^2}{u - a^2} + \frac{\mu m^2}{u - b^2} + \frac{\mu n}{u - c^2} = 0.$$

or $$\frac{l^2}{u - a^2} + \frac{m^2}{u - b^2} + \frac{n}{u - c^2} = 0. \quad ...(10)$$

These equation (10) gives the required maximum or minimum values of u.

Example 25(d):

If $u = x^2 + y^2 + z^2$,

where $ax^2 + by^2 + cz^2 + 2fy + 2gzx + 2hxy = 1$, find the maximum or minimum values of u.

Solution:

We have $u = x^2 + y^2 + z^2$, ...(1)

where the variables x, y, z are connected by the relation

$$ax^2 + by^2 + cz^2 + 2fyz + 2gzx + 2hxy = 1. \quad ...(2)$$

For a maximum or a minimum of u, we have

$$du = 0$$

$$\Rightarrow \quad 2x\,dx + 2y\,dy + 2z\,dz = 0$$

$$x\,dx + y\,dy + z\,dz = 0. \quad ...(3)$$

Also differentiating the given relation (2), we get

$$2ax\,dx + 2by\,dy + 2cz\,dz + 2fy\,dz + 2fz\,dy + 2gz\,dx + 2gx\,dz + 2\,hx\,dy + 2hy\,dx = 0$$

or $(ax + hy + gz)\,dx + (hx + by + fz)\,dy + (gx + fy + cz)\,dz = 0.$ ...(4)

Multiplying (3) by 1, (4) by λ and adding, and then equating the coefficients of dx, dy, dz to zero, we have

$$x + \lambda\,(ax + hy + gz) = 0, \quad ...(5)$$

$$y + \lambda\,(hx + by + fz) = 0, \quad ...(6)$$

and $\quad z + \lambda\,(gx + fy + cz) = 0.$ ...(7)

Multiplying (5) by x, (6) by y, (7) by z and adding, we get

$$x^2 + y^2 + z^2 + \lambda\,(ax^2 + by^2 + cz^2 + 2fyz + 2gzx + 2hxy) = 0$$

or $\quad u + \lambda.1 = 0$, using eqns. (1) and (2).

$\therefore \quad \lambda = -u.$

Hence from (5), we have

$$x - u\,(ax + hy + gz) = 0$$

or $\quad x(1 - au) - huy - guz = 0$

or $\quad \left(a - \frac{1}{u}\right)x + hy + gz = 0.$ ...(8)

Similarly from (6) and (7), we have

$$hx + \left(b - \frac{1}{u}\right)y + fz = 0, \quad ...(9)$$

and $\quad gx + fy + \left(c - \frac{1}{u}\right)z = 0.$ ...(10)

Eliminating x, y, z from (8), (9), (10), we get

$$\begin{vmatrix} a-(1/u) & h & g \\ h & b-(1/u) & f \\ g & f & c-(1/u) \end{vmatrix} = 0 \quad ...(1)$$

Hence the required maximum or minimum values of u are the roots of the equation (11).

Example 25(e):

Find the minimum value of x + y + z, subject to the condition

$$\left(\frac{a}{x}\right) + \left(\frac{b}{y}\right) + \left(\frac{c}{z}\right) = 1.$$

Solution:

Let $\quad u = x + y + z,$ ...(1)

where the variables x, y, z are connected by the relation

$$\frac{a}{x} + \frac{b}{y} + \frac{c}{z} = 1. \quad ...(2)$$

For a maximum or a minimum of u, we have

$$du = 0 \Rightarrow dx + dy + dz = 0. \quad ...(3)$$

Also differentiating the given equation (2), we get

$$-\frac{a}{x^2}dx-\frac{b}{y^2}dy-\frac{c}{z^2}dz=0. \qquad ...(4)$$

Multiplying (3) by 1 and (4) by λ, and adding and then equating the coefficients of dx, dy, dz to zero, we get

$$1-\frac{\lambda a}{x^2}=0,\ 1-\frac{\lambda b}{y^2}=0,\ 1-\frac{\lambda c}{z^2}=0.$$

From these, we get $x=\sqrt{(\lambda a)}$, $y=\sqrt{(\lambda b)}$, $z=\sqrt{(\lambda c)}$.

Putting these values of x, y, z in (2), we get

$$\frac{a}{\sqrt{(\lambda a)}}+\frac{b}{\sqrt{(\lambda b)}}+\frac{c}{\sqrt{(\lambda c)}}=1 \text{ or } \frac{1}{\sqrt{\lambda}}\left(\sqrt{a}+\sqrt{b}+\sqrt{c}\right)=1$$

or $\quad \sqrt{l}=\sqrt{a}+\sqrt{b}+\sqrt{c}\cdot$

$\therefore$ u is stationary when

$$\sqrt{l}=\sqrt{a}+\sqrt{b}+\sqrt{c},\ y=\sqrt{b}\left(\sqrt{a}+\sqrt{b}+\sqrt{c}\right),$$

$$z=\sqrt{c}\left(\sqrt{a}+\sqrt{b}+\sqrt{c}\right).$$

Let us regard x and y as independent variables and z as a function of x and y given by (2).

From (1), we have

$$\frac{\partial u}{\partial x}=1+\frac{\partial z}{\partial x}.$$

Differentiating (2) partially w.r.t. x taking y as constant, we get

$$-\frac{a}{x^2}-\frac{c}{z^2}\frac{\partial z}{\partial x}=0 \text{ or } \frac{\partial z}{\partial x}=-\frac{az^2}{cx^2}.$$

$$\therefore \qquad \frac{\partial u}{\partial x}=1-\frac{az^2}{cx^2},$$

so that $\dfrac{\partial^2 u}{\partial x^2}=\dfrac{2az^2}{cx^3}-\dfrac{2az}{cx^2}\dfrac{\partial z}{\partial x}=\dfrac{2az^2}{cx^3}+\dfrac{2az}{cx^2}\cdot\dfrac{az^2}{cx^2},$

which is positive for the values of x, y, z found above.

Hence, u is minimum at the stationary point found above and the minimum value of u

$$=\sqrt{a}\left(\sqrt{a}+\sqrt{b}+\sqrt{c}\right)+\sqrt{b}\left(\sqrt{a}+\sqrt{b}+\sqrt{c}\right)+\sqrt{c}\left(\sqrt{a}+\sqrt{b}+\sqrt{c}\right)$$

$$=\left(\sqrt{a}+\sqrt{b}+\sqrt{c}\right)^2.$$

Example 26:

Find the maxima and minima of $x^2 + y^2$ *subject to the condition* $ax^2 + 2hxy + by^2 = 1$.

Solution:

Let $u = x^2 + y^2$, ...(1)

where the variables x and y are connected by relation

$$ax^2 + 2hxy + by^2 = 1. \quad ...(2)$$

For a maximum or a minimum of u, we have du = 0

$\Rightarrow \quad 2xdx + 2ydy = 0$

$\Rightarrow \quad xdx + ydy = 0.$ ...(3)

Also differentiating the given relation (2), get

$$2axdx + 2hxdy + 2hydx + 2bydy = 0$$

or $(ax + hy)\, dx + (hx + by)\, dy = 0$ ....(4)

Multiplying (3) by 1, (4) by λ and adding, and then equating the coefficients of dx, dy to zero, we have

$$x + \lambda\,(ax + hy) = 0 \quad ...(5)$$

and $y + \lambda\,(hx + by) = 0$ ...(6)

Multiplying (5) by x, (6) by y and adding, we get

$$x^2 + y^2 + \lambda\,(ax^2 + 2hxy + by^2) = 0$$

or $u + \lambda\,.\,1 = 0,$ using (1) and (2).

$\therefore \quad \lambda = -u.$

Hence from (5), we have

$$x - u\,(ax + hy) = 0 \text{ or } x\,(1 - au) - huy = 0$$

or $\left(a - \frac{1}{u}\right)x + hy + 0$ ...(7)

Similarly from (6), we have

$$hx + \left(b - \frac{1}{u}\right)y = 0 \quad ...(8)$$

Eliminating x and y from (7) and (8), we get

$$\begin{vmatrix} a - \frac{1}{u} & h \\ h & b - \frac{1}{u} \end{vmatrix} = 0 \text{ or } \left(a - \frac{1}{u}\right)\left(b - \frac{1}{u}\right) = h^2 \quad ...(9)$$

Hence the required maximum or minimum values of $u = x^2 + y^2$ are the roots of the equation (9).

Example 27:

Find the maximum and minimum values of

$$\frac{x^2}{a^4}+\frac{y^2}{b^4}+\frac{z^2}{c^4},$$

when $lx + my + nz = 0$ *and* $\frac{x^2}{a^2}+\frac{y^2}{b^2}+\frac{z^2}{c^2}=1.$

Interpret the result geometrically.

Solution:

Let $u=\frac{x^2}{a^4}+\frac{y^2}{b^4}+\frac{z^2}{c^4}$. Then for a maximum or a minimum, of u, we have

$$du = 0$$

$$\Rightarrow \quad \frac{x}{a^4}dx+\frac{y}{b^4}dy+\frac{z}{c^4}dz=0 \qquad ...(1)$$

Also differentiating the two given equations connecting the variables x, y and z, we get

$$l dx + m\,dy + n\,dz = 0, \qquad ...(2)$$

and
$$\frac{x}{a^2}dx+\frac{y}{b^2}dy+\frac{z}{c^2}dz=0 \qquad ...(3)$$

Multiplying (1), (2) and (3) by 1, λ and μ respectively and adding, and then equating to zero the coefficients of dx, dy and dz, we get

$$\frac{x}{a^4}+\lambda l+\mu\frac{x}{a^2}=0, \qquad ...(4)$$

$$\frac{y}{b^4}+\lambda m+\mu\frac{y}{b^2}=0, \qquad ...(5)$$

and
$$\frac{z}{c^4}+\lambda n+\mu\frac{z}{c^2}=0. \qquad ...(6)$$

Multiplying the equations (4), (5) and (6) by x, y and z respectively and adding, we get

$$u + \lambda.0 + \mu.1 = 0 \text{ or } \mu = -u.$$

Putting $\mu = -u$ in (4), we get

$$\frac{x}{a^4}+\lambda l-\frac{xu}{a^2}=0, \text{ or } \frac{x}{a^2}\left\{u-\frac{1}{a^2}\right\}=\lambda l, \text{ or } x=\frac{\lambda u}{a^2u-1}.$$

Similarly from (5) and (6), we get

$$y=\frac{\lambda mb^4}{b^2u-1} \text{ and } z=\frac{\lambda nc^4}{c^2u-1}.$$

Substituting these values of x, y, z in $lx + my + nz = 0$, we get

$$\frac{l^2a^4}{a^2u-1}+\frac{m^2b^4}{b^2u-1}+\frac{n^2c^4}{c^2u-1}=0. \qquad ...(7)$$

The equation (7) gives the required maximum or minimum values of u.

Geometrical Interpretation: The equation of the tangent plane to the ellipsoid $\frac{x^2}{a^2}+\frac{y^2}{b^2}+\frac{z^2}{c^2}=1$ at any point (x, y, z) on it is

$$\frac{Xx}{a^2}+\frac{Yy}{b^2}+\frac{Zz}{c^2}=1. \qquad ...(8)$$

If p be the length of the perpendicular from origin which is also the centre of the ellipsoid to the tangent plane (1), then

$$p^2=\frac{1}{\frac{x^2}{a^4}+\frac{y^2}{b^4}+\frac{z^2}{c^4}}.$$

If the point (x, y, z) on the ellipsoid also lies on the given plane $lx + my + nz = 0$, the problem consists of finding out the maximum or minimum values of the perpendicular distance from the origin to the tangent planes to the ellipsoid at the points common to the plane $lx + my + nz = 0$ and the ellipsoid.

Example 28(a0:

Find the maximum value of $x^m y^n z^p$ subject to the condition

$$x + y + z = a.$$

Solution:

Let $u = x^m y^n z^p$, ...(1)

where the variables x, y, z are connected by the relation

$$x + y + z = a.$$

From (1), $\log u = m \log x + n \log y + p \log z$.

$$\therefore \quad \frac{1}{u}du=\frac{m}{x}dx+\frac{n}{y}dy+\frac{p}{z}dz.$$

For a maximum or a minimum of u, we have

$$du = 0$$

$$\Rightarrow \quad \frac{m}{x}dx+\frac{n}{y}dy+\frac{p}{y}dz=0. \qquad ...(3)$$

Differentiating the given equation (2) connecting the variables x, y and z, we get

$$dx + dy + dz = 0 \qquad ...(4)$$

Multiplying (3) by 1 and (4) by λ, and adding and then equating the coefficients of dx, dy, dz to zero, we get

$$\frac{m}{x}+\lambda=0,\ \frac{n}{y}+\lambda=0,\ \frac{p}{z}+\lambda=0.$$

From these, we get $x=-\frac{m}{\lambda},\ y=0\frac{n}{\lambda},\ z=-\frac{p}{\lambda}.$

Putting these values of x, y, z in x + y + z = a, we get

$$=\left(\frac{m}{\lambda}+\frac{n}{\lambda}+\frac{p}{\lambda}\right)=a$$

or $$-\frac{1}{\lambda}(m+n+p)=a$$

or $$-\frac{1}{\lambda}=\frac{a}{m+n+p}.$$

∴ u is stationary when

$$x=\frac{am}{m+n+p},\ y=\frac{an}{m+n+p},\ z=\frac{ap}{m+n+p}.$$

Let us now find the nature of this stationary value of u.

Since the variables x, y and z are connected by the relation (2), only two of them may be regarded as independent.

Let us regard x and y as independent variables and z as function of x and y given by (2).

From (1), we have

$$\log u = m\log x + n\log y + p\log z.$$

$$\therefore\quad \frac{1}{u}\frac{\partial u}{\partial x}=\frac{m}{x}+\frac{p}{z}\frac{\partial z}{\partial x}.$$

Differentiating (2) partially w.r.t. x taking y as constant, we get

$$1=\left(\frac{\partial z}{\partial x}\right)=0 \text{ or } \frac{\partial z}{\partial x}=-1.$$

$$\therefore\quad \frac{1}{u}\frac{\partial u}{\partial x}=\frac{m}{x}-\frac{p}{z},$$

so that $$\frac{1}{u}\frac{\partial^2 u}{\partial x^2}-\frac{1}{u^2}\left(\frac{\partial u}{\partial x}\right)^2=\frac{m}{x^2}+\frac{p}{z^2}\frac{\partial z}{\partial x}=-\frac{m}{x^2}-\frac{p}{z^2}.$$

But at the stationary point, we have $\frac{\partial u}{\partial x}=0.$

∴ at the stationary point found above, we have

$$\frac{\partial^2 u}{\partial x^2} = -u\left[\frac{m}{x^2}+\frac{p}{z^2}\right] = -x^m y^n z^p\left[\frac{m}{x^2}+\frac{p}{z^2}\right],$$

which is –ive for the values of x, y, z found above.

Hence at the stationary point found above the value of u is maximum and this maximum value

$$=\left(\frac{am}{m+n+p}\right)^m\left(\frac{an}{m+n+p}\right)^n\left(\frac{ap}{m+n+p}\right)^p = \frac{a^{m+n+p}m^m n^n p^p}{(m+n+p)^{m+n+p}}.$$

Example 28(b):

Find the maximum or minimum value of $x^p y^q z^r$ subject to the condition

$$ax + by + cz = p + q + r.$$

Solution:

For complete solution of this question proceed as in solved example.

Let $u = x^p y^q z^r$, ...(1)

where the variables x, y, z are connected by the relation

$$ax + by + cz = p + q + r. \qquad ...(2)$$

From (1), $\log u = p \log x + q \log y + r \log z$.

$$\therefore \quad \frac{1}{u}du = \frac{p}{x}dx + \frac{q}{y}dy + \frac{r}{z}dz.$$

For a maximum or a minimum of u, we have du = 0

$$\Rightarrow \quad \frac{p}{x}dx + \frac{q}{y}dy + \frac{r}{z}dz = 0. \qquad ...(3)$$

Also differentiating the given equation (2), we get

$$adx + b\,dy + c\,dz = 0. \qquad ...(4)$$

Multiplying eqns. (3) by 1 and (4) by λ, and adding and then equating the coefficients of dx, dy, dz to zero, we get

$$\frac{p}{x}+\lambda a = 0,\ \frac{q}{y}+\lambda b = 0,\ \frac{r}{z}+\lambda c = 0.$$

From these, we get $x = -\frac{p}{\lambda a},\ y = -\frac{q}{\lambda b},\ z = -\frac{r}{\lambda c}$.

Putting these values of x, y, z in (2), we get

$$-\left(\frac{p}{\lambda}+\frac{q}{\lambda}+\frac{r}{\lambda}\right) = p+q+r$$

or

$$-\frac{1}{\lambda}(p+q+r) = p+q+r$$

or $\lambda = -1$.

$\therefore$ u is stationary when $x = \frac{p}{a},\ y = \frac{q}{b},\ z = \frac{r}{c}$.

Now regard x and y as independent variables and z as a function of x and y given by (2).

From (1), we have

$\log u = p \log x + q \log y + r \log z$.

$$\therefore \quad \frac{1}{u}\frac{\partial u}{\partial x} = \frac{p}{x} + \frac{r}{z}\frac{\partial z}{\partial x}.$$

Differentiating (2) partially w.r.t. x taking y as constant, we get

$$a + c\left(\frac{\partial z}{\partial x}\right) = 0 \text{ or } \frac{\partial z}{\partial x} = -\frac{a}{c}.$$

$$\therefore \quad \frac{1}{u}\frac{\partial u}{\partial x} = \frac{p}{x} - \frac{r}{z}\cdot\frac{a}{c},$$

$$\text{so that } \frac{1}{u}\frac{\partial^2 u}{\partial x^2} - \frac{1}{u^2}\left(\frac{\partial u}{\partial x}\right)^2 = -\frac{p}{x^2} + \frac{a}{c}\cdot\frac{r}{z^2}\cdot\frac{\partial z}{\partial x} = -\frac{p}{x^2} - \frac{a^2}{c^2}\cdot\frac{r}{z^2}.$$

But at the stationary point, we have $\frac{\partial u}{\partial x} = 0$.

$\therefore$ at the stationary point found above, we have

$$\frac{\partial^2 u}{\partial x^2} = -u\left[\frac{p}{x^2} + \frac{z^2}{c^2}\cdot\frac{r}{z^2}\right] = -x^p y^q z^r\left[\frac{p}{x^2} + \frac{a^2}{c^2}\cdot\frac{r}{z^2}\right],$$

which is –ive for the values of x, y, z found above.

Hence u is maximum at the stationary point found above and this maximum value of $u = \left(\frac{p}{a}\right)^p \cdot \left(\frac{q}{b}\right)^q \cdot \left(\frac{r}{c}\right)^r$.

Example 28(c):

Find the maximum or minimum value of $x^p p^q z^r$ *subject to the condition*

$$\frac{a}{x} + \frac{b}{y} + \frac{c}{z} = 1.$$

Solution:

Let $u = x^p y^q z^r$, ...(1)

where the variables x, y, z are connected by the relation

$$\left(\frac{a}{x}\right) + \left(\frac{b}{y}\right) + \left(\frac{c}{z}\right) = 1. \qquad ...(2)$$

From (1), log u = p log x + q log y + r log z.

$$\therefore \qquad \frac{1}{u}du = \frac{p}{x}dx + \frac{q}{y}dy + \frac{r}{z}dz.$$

For a maximum or a minimum of u, we have du = 0.

$$\therefore \qquad \frac{p}{x}dx + \frac{q}{y}dy + \frac{r}{z}dz = 0. \qquad ...(3)$$

Also differentiating the given equation (2), we get

$$-\frac{a}{x^2}dx - \frac{b}{y^2}dy - \frac{c}{z^2}dz = 0. \qquad ...(4)$$

Multiplying (3) by 1 and (4) λ, and adding and then equating the coefficients of dx, dy, dz to zero, we get

$$\frac{p}{q} - \frac{\lambda a}{x^2} = 0, \frac{q}{y} - \frac{\lambda b}{y^2} = 0, \frac{r}{z} - \frac{\lambda c}{z^2} = 0.$$

From these, we get $x = \frac{a\lambda}{p}, y = \frac{b\lambda}{q}, z = \frac{c\lambda}{r}$.

Putting these values of x, y, z in (2), we get

$$\frac{p}{\lambda} + \frac{q}{\lambda} + \frac{r}{\lambda} = 1 \text{ or } \frac{1}{\lambda}(p + q + r) = 1 \text{ or } \lambda = p + q + r.$$

∴ u is stationary when

$$\frac{px}{a} = \frac{qy}{b} = \frac{rz}{c} = p + q + r.$$

Now, regard x and y as independent variables and z as a function of x and y given by (2).

From (1), we have

$$\log u = p \log x + q \log y + r \log z.$$

$$\therefore \qquad \frac{1}{u}\frac{\partial u}{\partial x} = \frac{p}{x} + \frac{r}{z}\frac{\partial z}{\partial x}.$$

Differentiating (2) partially w.r.t. x taking y as constant, we get

$$-\frac{a}{x^2} - \frac{c}{z^2}\frac{\partial z}{\partial x} = 0 \text{ or } \frac{\partial z}{\partial x} = -\frac{az^2}{cx^2}.$$

$$\therefore \qquad \frac{1}{u}\frac{\partial u}{\partial x} = \frac{p}{x} \cdot \frac{az^2}{cx^2} = \frac{p}{x} - \frac{arz}{cx^2},$$

so that $\frac{1}{u}\frac{\partial^2 u}{\partial x^2} - \frac{1}{u^2}\left(\frac{\partial u}{\partial x}\right)^2 = -\frac{p}{x^2} + \frac{2arz}{cx^3} - \frac{ar}{cx^2}\frac{\partial z}{\partial x}$

$$= -\frac{p}{x^2} + \frac{2arz}{cx^3} + \frac{ar}{cx^2} \cdot \frac{az^2}{cx^2}.$$

But at the stationary point, we have $\frac{\partial y}{\partial x} = 0$.

$\therefore$ at the stationary point found above, we have

$$\frac{\partial^2 u}{\partial x^2} = u\left[-\frac{p}{x^2} - \frac{2arz}{cx^3} + \frac{a^2rz^2}{c^2x^4}\right]$$

$$= x^p y^q z^r\left[-p \cdot \frac{p^2}{a^2(p+q+r)^2} + \frac{2ar}{c} \cdot \frac{c(p+q+r)}{r}\right.$$

$$= x^p y^q z^r\left[-\frac{p^3}{a^2(p+q+r)^2} + \frac{2p^2}{a^2(p+q+r)^2} + \frac{p^4}{ra^2(p+q+r)^2}\right]$$

$$= x^p y^q z^r\left[\frac{p^3}{a^2(p+q+r)^2} + \frac{p^4}{ra^2(p+q+r)^2}\right],$$

which is +ive for the values of x, y, z found above.

Hence u is minimum at the stationary point given by

$$\frac{px}{a} = \frac{qy}{b} = \frac{rz}{c} = p+q+r.$$

Also the minimum value of u

$$= \left[\frac{a(p+q+r)}{p}\right]^p \left[\frac{b(p+q+r)}{q}\right]^q \left[\frac{c(p+q+r)}{r}\right]^r$$

$$= \frac{a^p b^q c^r}{p^p q^q r^r}(p+q+r)^{p+q+r}.$$

Example 29(a):

A rectangular box open at the top is to have a given capacity. Find the dimensions of the box requiring least material for its construction.

Solution:

Let the given capacity of the box be V, its three edges be x, y, z and its surface be S. Then

$$S = xy + 2yz + 2zx \quad \text{...(1)}$$

and $\quad xyz = V. \quad$ ...(2)

For a maximum or a minimum of S, we have

$$dS = (y + 2z)\,dx + (x + 2z)\,dy + 2(x + y)\,dz = 0. \quad \text{...(3)}$$

Also from (2), since V is constant, we have

$$yz\,dx + zx\,dy + xy\,dz = 0. \qquad ...(4)$$

Multiplying (3) by 1 and (4) by λ, and adding and then equating to zero the coefficients of dx, dy, dz, we get

$$(y + 2z) + \lambda yz = 0, \qquad ...(5)$$

and $$2x + 2y + \lambda xy = 0. \qquad ...(6)$$

Multiplying (5) by x, (6) by y and subtracting, we get

$$2zx - 2zy = 0$$

or $$2z(x - y) = 0, \text{ or } x = y.$$

[The root $z = 0$ is inadmissible because the depth of the box cannot be zero.]

Similarly, from the equations (6) and (7), we get $y = 2z$.

Hence the dimensions of the box for a stationary value of S are

$$x = y = 2z = (2V)^{1/3}, \text{ from (2).}$$

Let us now find the nature of this stationary value of S.

Regard x and y as independent variables and z as a function of x and y given by (2).

Then from (1), $\dfrac{\partial S}{\partial x} = y + 2y\dfrac{\partial z}{\partial x} + 2z + 2x\dfrac{\partial z}{\partial x}$.

Differentiating (2) partially w.r.t. x taking y as constant, we get

$$yz + xy\frac{\partial z}{\partial x} = 0 \; \textit{i.e.,} \; \frac{\partial z}{\partial x} = -\frac{z}{x}.$$

$$\therefore \quad \frac{\partial S}{\partial x} = y - \frac{2yz}{x} + 2z - 2z = y - \frac{2yz}{x}$$

and so $\dfrac{\partial^2 S}{\partial x^2} = \dfrac{2yz}{x^2} - \dfrac{2y}{x} \cdot \dfrac{\partial z}{\partial x} = \dfrac{2yz}{x^2} + \dfrac{2yz}{x^2} = \dfrac{4yz}{x^2} = 2$ at $x = y = 2z$.

Thus, at the stationary point $x = y = 2z = (2V)^{1/3}$, we have

$$r = \frac{\partial^2 S}{\partial x^2} = 2, \text{ which is positive.}$$

Similarly, we can find $s = \dfrac{\partial^2 S}{\partial x\, \partial y}$ and $s = \dfrac{\partial^2 S}{\partial x\, \partial y}$ at the stationary point $x = y = 2z = (2V)^{1/3}$ and can show that $rt - s^2$ is positive.

Since at the stationary point $x = y = 2z = (2V)^{1/3}$, $rt - s^2 > 0$ and $r > 0$, therefore the stationary value of S at this point is a minimum.

Hence the dimensions of the box requiring least material for its construction are given by $x = y = 2z = (2V)^{1/3}$.

Example 29(b):

Find the minimum value of $x^4 + y^4 + z^4$, where $xyz = c^3$.

Solution:

Let $u = x^4 + y^4 + z^4$, ...(1)

where the variables x, y, z are connected by the relation

$$xyz = c^3. \quad ...(2)$$

For a maximum or a minimum of u, we have

$$du = 0 \Rightarrow 4x^3\,dx + 4y^3\,dy + 4z^3\,dz = 0$$

$$\Rightarrow \quad x^3\,dx + y^3\,dy + z^3\,dz = 0. \quad ...(3)$$

Also from the given relation (2), we have

$$\log x + \log y + \log z = \log c^3.$$

Differentiating this, we get

$$\left(\frac{1}{x}\right)dx + \left(\frac{1}{y}\right)dy + \left(\frac{1}{z}\right)dz = 0. \quad ...(4)$$

Multiplying eqns. (3) by 1 and (4) by λ, and adding and then equating the coefficients of dx, dy, dz to zero, we get

$$x^3 + \frac{\lambda}{x} = 0,\ y^3 + \frac{\lambda}{y} = 0,\ z^3 + \frac{\lambda}{z} = 0.$$

From these, we get $x^4 = y^4 = z^4 = -\lambda$.

Now from (2), $x^4y^4z^4 = c^{12}$.

$\therefore \quad -\lambda^3 = c^{12}$ or $\lambda = -c^4$.

$\therefore$ u is stationary when $x^4 = y^4 = z^4 = c^4$ *i.e.*, when $x = y = z = c$. Now, regard x and y as independent variables and z as a function of x and y given by (2).

From (1), we have $\dfrac{\partial u}{\partial x} = 4x^3 + 4z^3\dfrac{\partial z}{\partial x}$.

Now from (2), we have $\log x + \log y + \log z = \log c^3$.

Differentiating this partially w.r.t. x taking y as constant, we get

$$\frac{1}{x} + \frac{1}{z}\frac{\partial z}{\partial x} = 0 \text{ or } \frac{\partial z}{\partial x} = -\frac{z}{x}.$$

$$\therefore \quad \frac{\partial u}{\partial x} = 4x^3 - 4z^3\cdot\frac{z}{x} = 4x^3 - 4\frac{z^4}{x},$$

so that $$\frac{\partial^2 u}{\partial x^2} = 12x^2 + \frac{4z^4}{x^2} - \frac{16}{x}z^3\frac{\partial z}{\partial x}$$

$$= 12x^2 + \frac{4z^4}{x^2} - \frac{16z^3}{x}\left(-\frac{z}{x}\right) = 12x^2 + \frac{4z^4}{x^2} + \frac{16z^4}{x^2}.$$

At the stationary point (c, c, c) found above, we have

$$\frac{\partial^2 u}{\partial x^2} = 12c^2 + 4c^2 + 16c^2 = 32c^2, \text{ which is +ive.}$$

$\therefore$ u is minimum at the point x = y = z = c and the minimum value of

$$u = c^4 + c^4 + c^4 = 3c^4.$$

Example 29(c):

Prove that the rectangular solid of maximum volume which can be inscribed in a sphere is a cube.

Solution:

Referred to the centre as origin, let the equation of the sphere

$$x^2 + y^2 + z^2 = a^2.$$

Let (x, y, z) denote the coordinates of that vertex of the rectangular parallelopiped inscribed in the sphere which lies in the positive octant and let V denote the volume of the rectangular parallelopiped. Then, we have to find the maximum value of

$$V = 8xyz \qquad ...(1)$$

subject to the condition

$$x^2 + y^2 + z^2 = a^2. \qquad ...(2)$$

From (1), log V = log 8 + log x + log y + log z.

$$\therefore \quad \frac{1}{V}dV = \frac{1}{x}dx + \frac{1}{y}dy + \frac{1}{z}dz.$$

For a maximum or a minimum of V, we must have dV = 0

$$\Rightarrow \quad \frac{1}{x}dx + \frac{1}{y}dy + \frac{1}{z} = 0. \qquad ...(3)$$

Also differentiating (2), we have

$$x\,dx + y\,dy + z\,dz = 0. \qquad ...(4)$$

Multiplying (3) by 1 and (4) by λ, and adding and then equating to zero the coefficients of dx, dy, dz, we get

$$\frac{1}{x} + \lambda x = 0, \quad \frac{1}{y} + \lambda y = 0, \quad \frac{1}{z} + \lambda z = 0$$

or $\quad -\frac{1}{\lambda} = x^2 = y^2 = z^2$ or x = y = z.

Thus V is stationary when $x = y = z = \frac{a}{\sqrt{3}}$, from (2).

The lengths of the edges of the rectangular parallelopiped is a cube.

Now regard x and y as independent variables and z as a function of x and y given by (2).

From (1), log V = log 8 + log x + log y + log z.

$$\therefore \quad \frac{1}{V}\frac{\partial V}{\partial x}=\frac{1}{x}+\frac{1}{z}\cdot\frac{\partial z}{\partial x}.$$

Differentiating (2) partially w.r.t. x taking y as constant, we get

$$2x+2z\left(\frac{\partial z}{\partial x}\right)=0 \text{ or } \frac{\partial z}{\partial x}=-\frac{x}{z}.$$

$$\therefore \quad \frac{1}{V}\frac{\partial V}{\partial x}=\frac{1}{x}+\frac{1}{z}\cdot\frac{-x}{z}=\frac{1}{x}-\frac{x}{z^2},$$

so that $\frac{1}{V}\frac{\partial^2 V}{\partial x^2}-\frac{1}{\sqrt{2}}\left(\frac{\partial V}{\partial x}\right)^2=-\frac{1}{x^2}-\frac{1}{z^2}+\frac{2x}{z^3}\frac{\partial z}{\partial x}=-\frac{1}{x^2}-\frac{1}{z^2}-\frac{2x^2}{z^4}$

But at the stationary point, we have $\frac{\partial V}{\partial x}=0$.

∴ at the stationary point found above, we have

$$\frac{\partial^2 V}{\partial x^2}=-V\left[\frac{1}{x^2}+\frac{1}{z^2}+\frac{2x^2}{z^4}\right]=-8xyz\left[\frac{1}{x^2}+\frac{1}{z^2}+\frac{2x^2}{z^4}\right],$$

which is –ive when $x = y = z = \frac{a}{\sqrt{3}}$.

Thus V is maximum when $x = y = z = \frac{a}{\sqrt{3}}$.

Hence the rectangular solid of maximum volume inscribed in a sphere is a cube.

Example 29(d):

Find the maximum value of u, when

$$u = x^2y^3z^4 \text{ and } 2x + 3y + 4z = a.$$

Solution:

Let $u = x^2y^3z^4$, ...(1)

where the variables x, y, z are connected by the relation

$2x + 3y + 4z = a$. ...(2)

From (1), log u = 2 log x + 3 log y + 4 log z.

$$\therefore \quad \frac{1}{u}du=\frac{2}{x}dx+\frac{3}{y}dy+\frac{4}{z}dz.$$

For a maximum or a minimum of u, we have

$$du = 0 \Rightarrow \left(\frac{2}{x}\right)dx+\left(\frac{3}{y}\right)dy+\left(\frac{4}{z}\right)dz=0. \quad ...(3)$$

Differentiating the given equation (2), we have

$$2dx + 3dy + 4dz = 0. \quad \text{...(4)}$$

Multiplying eqns. (3) by 1 and (4) by λ, and adding and then equating the coefficients of dx, dy, dz to zero, we get

$$\frac{2}{x}+2\lambda=0,\ \frac{3}{y}+3\lambda=0,\ \frac{4}{z}+4\lambda=0.$$

From these, get $x=-\frac{1}{\lambda},\ y=-\frac{1}{\lambda},\ z=-\frac{1}{\lambda}.$

Putting these values of x, y, z in (2), we get

$$-\frac{2}{\lambda}-\frac{3}{\lambda}-\frac{4}{\lambda}=a \text{ or } -\frac{9}{\lambda}=a \text{ or } \lambda=-\frac{9}{a}.$$

∴ u is stationary when $x = y = z = \frac{a}{9}.$

Now regard x and y as independent variables and z as a function of x and y given by (2).

From (1), we have

$$\log u = 2 \log x + 3 \log y + 4 \log z.$$

$$\therefore \quad \frac{1}{u}\frac{\partial u}{\partial x}=\frac{2}{x}+\frac{4}{z}\frac{\partial z}{\partial x}.$$

Differentiating (2) partially w.r.t. x taking y as constant, we get

$$2+4\left(\frac{\partial z}{\partial x}\right)=0 \text{ or } \frac{\partial z}{\partial x}=-\frac{1}{2}.$$

$$\therefore \quad \frac{1}{u}\frac{\partial u}{\partial x}=\frac{2}{x}+\frac{4}{2}\cdot\left(-\frac{1}{2}\right)=\frac{2}{x}-\frac{2}{z}.$$

so that $\frac{1}{u}\frac{\partial^2 u}{\partial x^2}-\frac{1}{u^2}\left(\frac{\partial u}{\partial x}\right)^2=-\frac{2}{x^2}+\frac{2}{z^2}\frac{\partial z}{\partial x}=-\frac{2}{x^2}-\frac{1}{z^2}$

But at the stationary point, we have $\frac{\partial u}{\partial x}=0.$

∴ at the stationary point found above, we have

$$\frac{\partial^2 u}{\partial x^2}=-u\left(\frac{2}{x^2}+\frac{1}{z^2}\right)=-x^2y^3z^4\left(\frac{2}{x^2}+\frac{1}{z^2}\right),$$

which is –ive for $x = y = z = \frac{a}{9}.$

Hence at the stationary point $x = y = z = \frac{a}{9}$, u is maximum and the maximum value of u

$$=\left(\frac{a}{9}\right)^2\left(\frac{a}{9}\right)^3\left(\frac{a}{9}\right)^4=\left(\frac{a}{9}\right)^9.$$

Example 30:

Show that the volume of the greatest rectangular parallelopiped that can be inscribed in the ellipsoid

$$\frac{x^2}{a^2}+\frac{y^2}{b^2}+\frac{z^2}{c^2}=1$$

is $\frac{8abc}{3\sqrt{3}}$.

Solution:

Let (x, y, z) denote the coordinates of the vertex of the rectangular parallelopiped which lies in the positive octant and let V denote its volume. Then, we have to find the maximum value of

$$V = 8xyz \qquad ...(1)$$

subject the condition

$$\frac{x^2}{a^2}+\frac{y^2}{b^2}+\frac{z^2}{c^2}=1. \qquad ...(2)$$

For a maximum or a minimum of V, we have

$$dV = 8yz\,dx + 8zx\,dy + 8xy\,dz = 0$$

i.e.,

$$yz\,dx + zx\,dy + xy\,dz = 0. \qquad ...(3)$$

Also differentiating (2), we get

$$\frac{2x}{a^2}dx+\frac{2y}{b^2}dy+\frac{2z}{c^2}dz=0$$

i.e.,

$$\frac{x}{a^2}dx+\frac{y}{b^2}dy+\frac{z}{c^2}dz=0. \qquad ...(4)$$

Multiplying (3) by 1 and (4) by λ, and adding and then equating the coefficients of dx, dy, dz to zero, we get

$$yz+\frac{\lambda x}{a^2}=0,\ zx+\frac{\lambda y}{b^2}=0 \text{ and } xy+\frac{\lambda z}{c^2}=0.$$

From these, we get

$$\frac{x}{a^2}=-\frac{yz}{\lambda},\ \frac{y}{b^2}=-\frac{zx}{\lambda},\ \frac{z}{c^2}=-\frac{xy}{\lambda}$$

or

$$\frac{x^2}{a^2}=-\frac{xyz}{\lambda},\ \frac{y^2}{b^2}=-\frac{xyz}{\lambda},\ \frac{z^2}{c^2}=-\frac{xyz}{\lambda}$$

or $$\frac{x^2}{a^2} = \frac{y^2}{b^2} = \frac{z^2}{c^2} = \frac{x^2/a^2 + y^2/b^2 + z^2/c^2}{3} = \frac{1}{3}, \text{ using (2)}$$

or $$x = \frac{a}{\sqrt{3}}, y = \frac{b}{\sqrt{3}}, z = \frac{c}{\sqrt{3}}.$$

Thus, V is stationary when $x = \frac{a}{\sqrt{3}}, y = \frac{b}{\sqrt{3}}, z = \frac{c}{\sqrt{3}}$.

Now regard x and y as independent variables and z as a function of x and y given by (2).

Then from (1), $\frac{\partial V}{\partial x} = 8yz + 8xy\frac{\partial z}{\partial x}$.

Differentiating (2) partially w.r.t. x taking y as constant, we get

$$\frac{2x}{a^2} + \frac{2z}{c^2}\frac{\partial z}{\partial x} = 0 \text{ or } \frac{\partial z}{\partial x} = -\frac{c^2 x}{a^2 z}.$$

$$\therefore \quad \frac{\partial V}{\partial x} = 8yz + 8xy \cdot \left(-\frac{c^2 x}{a^2 z}\right) = 8yz - \frac{8c^2 x^2 y}{a^2 z}$$

and so $$\frac{\partial^2 V}{\partial x^2} = 8y\frac{\partial z}{\partial x} - \frac{16c^2 xy}{a^2 z} + \frac{8c^2 x^2 y}{a^2 z^2} \cdot \frac{\partial z}{\partial x}$$

$$= 8y \cdot \left(\frac{-c^2 x}{a^2 z}\right) - \frac{16c^2 xy}{a^2 z} - \frac{8c^2 x^2 y}{a^2 z} \cdot \frac{c^2 x}{a^2 z},$$

which is –ive when $x = \frac{a}{\sqrt{3}}, y = \frac{b}{\sqrt{3}}, z = \frac{c}{\sqrt{3}}$.

Hence V is maximum when $x = \frac{a}{\sqrt{3}}, y = \frac{b}{\sqrt{3}}, z = \frac{c}{\sqrt{3}}$ and the maximum value of $V = \frac{8abc}{3\sqrt{3}}$.

Note: In complicated problems to show that whether the stationary value of a function is maximum or minimum, it will be sufficient to see whether the second partial differential coefficient of the function w.r.t. any of the independent variables is negative or positive.

Example 31:

Find the triangle of maximum area inscribed in a circle.

Solution:

Let x, y, z denote the angles of a triangle inscribed in a given circle of radius k. If u be the area of the triangle, then

$$u = \frac{1}{2}k^2 \ (\sin 2x + \sin 2y + \sin 2z), \qquad ...(1)$$

where the variables x, y, z are connected by the relation

$$x + y + z = \pi. \qquad ...(2)$$

From (1), $du = k^2 (\cos 2x\, dx + \cos 2y\, dy + \cos 2z\, dz)$.

For a maximum or a minimum of u, we must have du = 0

$$\Rightarrow \cos 2x\, dx + \cos 2y\, dy + \cos 2z\, dz = 0. \qquad ...(3)$$

Also differentiating (2), we have

$$dx + dy + dz = 0 \qquad ...(4)$$

Multiplying (3) by 1 and (4) by λ, and adding and then equating to zero the coefficients of dx, dy and dz, we get

$$\cos 2x + \lambda = 0,\ \cos 2y + \lambda = 0,\ \cos 2z + \lambda = 0$$

or $\quad -\lambda = \cos 2x = \cos 2y = \cos 2z$

or $\quad 1 - 2\sin^2 x = 1 - 2\sin^2 y = 1 - 2\sin^2 z$

or $\quad \sin^2 x = \sin^2 y = \sin^2 z$

or $\quad \sin x = \sin y = \sin^2 z$

or $\quad x = y = z = \frac{\pi}{3}$, from (2).

Thus, u is stationary when $x = y = z = \frac{\pi}{3}$ *i.e.*, the triangle is equilateral.

Now to show that the stationary value of u given by $x = y = z = \frac{\pi}{3}$ is maximum.

Let us regard x and y as independent variables and z as a function of x and y given by (2).

From eqn. (1), $\frac{\partial u}{\partial x} = k^2\left(\cos 2x + \cos 2z \frac{\partial z}{\partial x}\right)$.

Also differentiating (2) partially w.r.t. x taking y as a constant, we get

$$1 + \left(\frac{\partial z}{\partial x}\right) = 0$$

or $\quad \frac{\partial z}{\partial x} = -1.$

$\therefore \quad \frac{\partial u}{\partial x} = k^2(\cos 2x - \cos 2z),$

so that $\quad \frac{\partial^2 u}{\partial x^2} = k^2\left[-2\sin 2x + 2\sin 2z \cdot \frac{\partial z}{\partial x}\right]$

$$= -2k^2 (\sin 2x + \sin 2z).$$

Also $\dfrac{\partial^2 u}{\partial x\, \partial y} = k^2\left(2\sin 2z \dfrac{\partial z}{\partial y}\right) = -2k^2 \sin 2z.$

Hence putting $x = y = z = \dfrac{\pi}{3}$, we get

$$r = \frac{\partial^2 u}{\partial x^2} = -2k^2\left(\frac{\sqrt{3}}{2} + \frac{\sqrt{3}}{2}\right) = -2k^2\sqrt{3}.$$

$$s = \frac{\partial^2 u}{\partial x \partial y} = -2k^2 \cdot \frac{\sqrt{3}}{2} = -k^2\sqrt{3}.$$

Also by symmetry, $t = \dfrac{\partial^2 u}{\partial y^2} = -2k^2\sqrt{3}.$

$\therefore$ at the stationary point $x = y = z = \dfrac{\pi}{3}$, we have

$$rt - s^2 = 12k^4 - 3k^4 = 9k^4 \text{ which is } > 0$$

and $r = -2k^2\sqrt{3}$ which is –ive.

$\therefore$ at the stationary point $x = y = z = \dfrac{\pi}{3}$, u is maximum.

Hence the triangle of maximum area inscribed in a circle is equilateral.

Example 32:

Show that if the perimeter of a triangle is constant, its area is a maximum when it is equilateral.

Solution:

Let a, b, c denote the sides of a triangle, 2s is constant perimeter and u its area.

Then $u^2 = s(s-a)(s-b)(s-c)$, ...(1)

where the variables a, b, c are connected by the relation

$$a + b + c = 2s. \quad ...(2)$$

From (1), $2 \log u = \log s + \log(s-a) + \log(s-b) + \log(s-c)$.

$$\therefore \quad \frac{2}{u}du = -\frac{1}{s-a}da - \frac{1}{s-b}db - \frac{1}{s-c}dc.$$

For a maximum or a minimum of u, we must have du = 0

$$\Rightarrow \quad \frac{da}{s-a} + \frac{db}{s-a} + \frac{dc}{s-c} = 0. \quad ...(3)$$

Also differentiating (2), we have

$$da + db + dc = 0. \quad ...(4)$$

Multiplying (3) by 1 and (4) by λ, and adding and when equating to zero the coefficients of da.db and dc, we get

$$\frac{1}{s-a}+\lambda=0, \frac{1}{s-b}+\lambda=0, \frac{1}{s-c}+\lambda=0.$$

$$\therefore \qquad -\lambda=\frac{1}{s-a}=\frac{1}{s-b}=\frac{1}{s-c}$$

or $\qquad$ s – a = s – b = s -- c or a = b = c.

Hence, u is stationary when a = b = c *i.e.*, the triangle is equilateral.

Now, to show that the stationary value of u given by a = b = c is maximum.

Let us regard a and b as independent variables and c as a function of a and b given by (2).

From (1), differentiating logarithmically, we have

$$\frac{2}{u}\frac{\partial u}{\partial a}=-\frac{1}{s-a}-\frac{1}{s-c}\frac{\partial c}{\partial a}.$$

But from (2), $1+\left(\frac{\partial c}{\partial a}\right)=0$ or $\frac{\partial c}{\partial a}=-1$.

$$\therefore \qquad \frac{2}{u}\frac{\partial^2 u}{\partial a^2}-\frac{2}{u^2}\left(\frac{\partial u}{\partial a}\right)^2=-\frac{1}{(s-a)^2}+\frac{1}{(s-c)^2}\cdot\frac{\partial c}{\partial a}$$

$$=-\left[\frac{1}{(s-a)^2}+\frac{1}{(s-c)^2}\right].$$

But at the stationary point, we have $\frac{\partial u}{\partial a}=0$.

∴ at the stationary point found above, we have

$$\frac{2}{u}\frac{\partial^2 u}{\partial a^2}=-\left[\frac{1}{(s-a)^2}+\frac{1}{(s-c)^2}\right]$$

or $\qquad \frac{\partial^2 u}{\partial a^2}=-\frac{u}{2}\left[\frac{1}{(s-a)^2}+\frac{1}{(s-c)^2}\right]$, which is –ive.

Hence u is maximum when a = b = c *i.e.*, the area of the triangle is maximum when it is equilateral.

Example 33:

Given $u=\frac{5xyz}{(x+2y+4z)}$. *Find the values of x, y, z for which u is maximum subject to the condition xyz = 8.*

Solution:

We have $u = \frac{5xyz}{(x+2y+4z)}$ ...(1)

where the variables x, y, z are connected by the relation

$xyz = 8.$ ...(2)

From (1) and (2), we have $u = \frac{40}{(x+2y+4z)}$.

$\therefore \quad du = \frac{-40}{(x+2y+4z)^2}(dx + 2dy + 4dz)$.

For a maximum of a minimum of u, we have du = 0

$\Rightarrow \quad dx + 2dy + 4dz = 0.$ ...(3)

From (2), log x + log y + log z = log 8.

Differentiating this, we get

$$\left(\frac{1}{x}\right)dx + \left(\frac{1}{y}\right)dy + \left(\frac{1}{z}\right)dz = 0. \qquad ...(4)$$

Multiplying eqns. (3) by 1 and (4) by λ, and adding and then equating to zero the coefficients of dx, dy and dz, we get

$$1 + \left(\frac{\lambda}{x}\right) = 0,\ 2 + \left(\frac{\lambda}{y}\right) = 0,\ 4 + \left(\frac{\lambda}{z}\right) = 0.$$

From these, we get $x = -\lambda$, $y = -\frac{\lambda}{2}$, $z = -\frac{\lambda}{4}$.

Putting these values of x, y, z in (2), we get

$$-\frac{\lambda^3}{8} = 8 \text{ or } \lambda^3 = -64 \text{ or } \lambda = -4.$$

∴ u is stationary at the given by x = 4, y = 2, z = 1.

Now regard x and y as independent variables and z as a function of x and y given by (2).

We have $u = \frac{40}{(x+2y+4z)}$.

$$\therefore \quad \frac{\partial u}{\partial x} = -\frac{40}{(x+2y+4z)^2}\left[1 + 4\frac{\partial z}{\partial x}\right].$$

From (2), log x + log y + log z = log 8.

$$\therefore \quad \left(\frac{1}{x}\right) + \left(\frac{1}{z}\right)\left(\frac{\partial z}{\partial x}\right) = 0$$

or $\quad \dfrac{\partial z}{\partial x} = -\dfrac{z}{x}$.

$\therefore \quad \dfrac{\partial u}{\partial x} = -\dfrac{40}{(x+2y+4z)^2}\left[1-4\dfrac{z}{x}\right],$

so that $\dfrac{\partial^2 u}{\partial x^2} = \dfrac{80}{(x+2y+4z)^3}\left[1+4\dfrac{\partial z}{\partial x}\right]\left[1-4\dfrac{z}{x}\right]$

$$-\frac{40}{(x+2y+4z)^2}\left[\frac{4z}{x^2}-\frac{4}{x}\frac{\partial z}{\partial x}\right]$$

$$= \frac{80}{(x+2y+4z)^3}\left[1-\frac{4z}{x}\right]^2 - \frac{40}{(x+2y+4z)^2}\left[\frac{4z}{x^2}+\frac{4z}{x^2}\right].$$

$\therefore$ at the stationary point (4, 2, 1) found above, we have

$$\frac{\partial^2 u}{\partial x^2} = \frac{80}{12^3}[1-1]^2 - \frac{40}{144}\left[\frac{1}{4}+\frac{1}{4}\right], \text{ which is –ive.}$$

$\therefore$ u is maximum at the point given by x = 4, y = 2, z = 1.

Example 34:

Find the plane triangle ABC such that

$$u = \sin^m A \sin^n B \sin^p C$$

has maximum value.

Solution:

We have $u = \sin^m A \sin^n B \sin^p C$, ...(1)

where the variables A, B and C are connected by the relation

$$A + B + C = \pi. \quad ...(2)$$

From (1), log u = m log sin A + n log sin B + p log sin C.

$$\therefore \quad \frac{1}{u}du = m \cot A\, dA + n \cot B\, dB + p \cot C\, dC.$$

For a maximum or a minimum of u we must have du = 0

$$\Rightarrow \quad m \cot A\, dA + n \cot B\, dB + p \cot C\, dC = 0. \quad ...(3)$$

Also differentiating (2), we get

$$dA + dB + dC = 0. \quad ...(4)$$

Multiplying eqns. (3) by 1 and (4) by λ, and adding and then equating to zero the coefficients of dA.dB and dC, we get

$$m \cot A + \lambda = 0,\ n \cot B + \lambda = 0,\ p \cot C + \lambda = 0.$$

$$\therefore \quad -\lambda = m \cot A = n \cot B = p \cot C.$$

Hence u is stationary when A, B, C are given by

$$m \cot A = n \cot B = p \cot C$$

or
$$\frac{\tan A}{m} = \frac{\tan B}{n} = \frac{\tan C}{p}.$$

Now to show that the above stationary value of u is maximum.

Let us regard A and B as independent variables and C as a function of A and B given by (2).

From (1), log u = m log sin A + n log sin B + p log sin C.

$$\therefore \quad \frac{1}{u}\frac{\partial u}{\partial A} = m \cot A + p \cot C \cdot \frac{\partial C}{\partial A}.$$

But from (2), $1 + \left(\frac{\partial C}{\partial A}\right) = 0$ or $\frac{\partial C}{\partial A} = -1.$

$$\therefore \quad \frac{1}{u}\frac{\partial u}{\partial A} = m \cot A - p \cot C,$$

so that $\frac{1}{u}\frac{\partial^2 u}{\partial A^2} - \frac{1}{u^2}\left(\frac{\partial u}{\partial A}\right)^2 = -m \operatorname{cosec}^2 A + p \operatorname{cosec}^2 C \cdot \frac{\partial C}{\partial A}$

$$= -(m \operatorname{cosec}^2 A + p \operatorname{cosec}^2 C).$$

But at the stationary point found above, we have

$$\frac{\partial^2 u}{\partial A^2} = -u\left(m \operatorname{cosec}^2 A + p \operatorname{cosec}^2 C\right)$$

$$= -\sin^m A \sin^n B \sin^p C \,(m \operatorname{cosec}^2 A + p \operatorname{cosec}^2 C),$$

which is obviously –ive if A, B, C are the angles of a triangle.

Hence u is maximum when A, B, C are given by

$$\frac{\tan A}{m} = \frac{\tan B}{n} = \frac{\tan C}{p}.$$

4

Multi Variable Functions

INTRODUCTION

We have studied functions of a single independent variable such as:

$$y = f(x) \qquad \ldots(1)$$

In Economics, there are a number of cases where we have to deal with more than one independent variable. For example, we know that cost of a commodity produced is a function of both labour and material. i.e., if x denotes the labour and y denotes the material, then cost of a commodity is given by:

$$C = f(x, y) \qquad \ldots(2)$$

Again, consider demand for a commodity. It depends on price of the commodity, income of the consumer and prices of related goods. So, if x denotes the price of the commodity, y denotes the income of the consumer and z denotes the price of a related good, then demand for a commodity is:

$$q = f(x, y, z) \qquad \ldots(3)$$

In equation (2), x and y are considered to be independent variables and C, the dependent variable. There exists only one value of C in the range of f for each ordered pair of real numbers (x, y) in the domain of f. Here C is called a *multivariable function.* In equation (3) also, q is a multivariable function. The diagrammatic representation of a function of more than two variables is not possible in a 2-dimensional plane because it would require more than three dimensions.

As in the two-dimensional case, the relation, $z = f(x, y)$ is said to be a function of x and y over a region R, if for each pair of values of x and y in the region R, there is at least one value for z.

More over, $f(x, y)$ is said to be continuous at $x = a$, $y = b$, if:

(i) $f(a, b)$ exists

(ii) $\lim\limits_{\substack{x \to a \\ y \to b}} f(x, y)$ exists

(iii) $\lim\limits_{\substack{x \to a \\ y \to b}} f(x, y) = f(a, b)$,

no matter in what manner $x \to a$, and $y \to b$.

$f(x, y)$ is said to be *continuous* in a region of the xy-plane if it is continuous at every point of the region.

Now consider equation (2) again. A change in x or in y or in both will make a corresponding change in C. To measure the effect of a change in a single independent variable on C, we 'partially differentiate' C with respect to that single independent variable. Partial differentiation of C with respect to a single independent variable means, the differentiation of C with respect to that single independent variable, treating all other independent variables in C as constants.

EULER'S THEOREM

Euler's Theorem on homogeneous functions states that if $f(x, y)$ is a homogeneous function of degree n, then,

$$x\frac{\partial f}{\partial x}+y\frac{\partial f}{\partial y} = \text{n } f(x, y)$$

Proof:

Since $f(x, y)$ is a homogeneous function of degree n, it can be expressed in the form:

$$f(x, y) = x^{n} \varphi\left(\frac{y}{x}\right) \qquad \text{...(A)}$$

Differencing (A) partially with respect to x,

$$\frac{\partial f}{\partial x} = \frac{\partial}{\partial x}\left[x^{n}\varphi\left(\frac{y}{x}\right)\right]$$

$$= x^{n} \varphi'\left(\frac{y}{x}\right)\left(\frac{-y}{x^2}\right) + \text{n } x^{n-1} \varphi\left(\frac{y}{x}\right)$$

$$= - x^{n-2} y \varphi'\left(\frac{y}{x}\right) + \text{n } x^{n-1} \varphi\left(\frac{y}{x}\right)$$

$$\therefore \quad x\frac{\partial f}{\partial x} = - x^{n-1} y \varphi'\left(\frac{y}{x}\right) + \text{n } x^{n} \varphi\left(\frac{y}{x}\right) \qquad \text{...(1)}$$

Now, differentiating (A) partially with respect to y,

$$\frac{\partial f}{\partial y} = \frac{\partial}{\partial y}\left[x^n \varphi\left(\frac{y}{x}\right)\right]$$

$$= x^n \varphi'\left(\frac{y}{x}\right)\left(\frac{1}{x}\right) = x^{n-1} \varphi'\left(\frac{y}{x}\right)$$

$$\therefore \quad y\frac{\partial f}{\partial y} = - x^{n-1}\, y\, \varphi'\left(\frac{y}{x}\right) \qquad \ldots(2)$$

Adding (1) and (2),

$$x\frac{\partial f}{\partial x} + y\frac{\partial f}{\partial y} = - x^{n-1}\, y\, \varphi'\left(\frac{y}{x}\right) + n\, x^n\, \varphi\left(\frac{y}{x}\right) - x^{n-1}\, y\, \varphi'\left(\frac{y}{x}\right)$$

$$= n\, x^n\, \varphi\left(\frac{y}{x}\right) = n\, f(x, y)$$

Remark:

The result of the theorem can be extended to three variables as:
If $f(x, y, z)$ is a homogeneous function in x, y and z of degree n, then,

$$x\frac{\partial f}{\partial x} + y\frac{\partial f}{\partial y} + z\frac{\partial f}{\partial z} = n\, f(x, y, z).$$

Example:

Verify Euler's theorem for the function u = $\sqrt{x} + \sqrt{y}$

Solution:

First let us test the function for homogeneity.
Put x = tx and y = ty

Then $\quad u(tx, ty) = \sqrt{tx} + \sqrt{ty} = \sqrt{t}\left(\sqrt{x} + \sqrt{y}\right) = t^{1/2} u$

Hence u is homogeneous with degree ½.
Now differentiating u partially with respect to x,

$$\frac{\partial u}{\partial x} = \frac{1}{2\sqrt{x}}$$

$$\therefore \quad x\frac{\partial u}{\partial x} = \frac{x}{2\sqrt{x}} = \frac{\sqrt{x}}{2} \qquad \ldots(1)$$

From symmetry,

$$\therefore \qquad y\frac{\partial u}{\partial y} = \frac{\sqrt{y}}{2} \qquad \text{...(2)}$$

$$\text{Hence } x\frac{\partial u}{\partial x} + y\frac{\partial u}{\partial y} = \frac{\sqrt{x}}{2} + \frac{\sqrt{y}}{2} = \frac{1}{2}\left(\sqrt{x} + \sqrt{y}\right) = ½ \text{ u}$$

This verifies Euler's theorem.

SECOND ORDER PARTIAL DERIVATIVES

Given a function $z = f(x, y)$, the second order direct partial derivative signifies that z has been differentiated partially with respect to one of the independent variables x or y twice while the other independent variable has been held constant. Second order partial derivatives are of two kinds; one is *direct partial derivative* and the other is *cross partial derivative*. The second order direct partial derivative of z with respect to x is denoted by:

$$\frac{\partial^2 z}{\partial x^2} \text{ (or } z_{xx} \text{ or } f_{xx})$$

and with respect to y is denoted by:

$$\frac{\partial^2 z}{\partial y^2} \text{ (or } z_{yy} \text{ or } f_{yy}).$$

$$\text{i.e.,} \qquad \frac{\partial^2 z}{\partial x^2} = \frac{\partial}{\partial x}\left(\frac{\partial z}{\partial x}\right) = z_{xx} \text{ or } f_{xx} \text{ ; and}$$

$$\frac{\partial^2 z}{\partial y^2} = \frac{\partial}{\partial y}\left(\frac{\partial z}{\partial y}\right) = z_{yy} \text{ or } f_{yy}$$

Thus $\frac{\partial^2 z}{\partial x^2}$ measures the rate of change of $\frac{\partial z}{\partial x}$ with respect to x while y is held constant and $\frac{\partial^2 z}{\partial y^2}$ measures the rate of change of $\frac{\partial z}{\partial y}$ with respect to y while x is held constant.

The cross partial derivatives of z indicate that z has been partially differentiated with respect to one independent variable and then that derivative has in turn been partially differentiated with respect to the other independent variable. The cross partial derivative of z, firstly with respect to x and then with respect to y is denoted by:

$$\frac{\partial^2 z}{\partial y \partial x} \text{ (or } z_{yx} \text{ or } f_{yx}) \text{ ;}$$

and firstly with respect to y and then with respect to x is denoted by:

$$\frac{\partial^2 z}{\partial x \partial y} \text{ (or } z_{xy} \text{ or } f_{xy}).$$

i.e.,
$$\frac{\partial^2 z}{\partial x \partial y} = \frac{\partial}{\partial y}\left(\frac{\partial z}{\partial x}\right) = z_{yx} \text{ or } f_{yx} \text{; and}$$

$$\frac{\partial^2 z}{\partial x \partial y} = \frac{\partial}{\partial x}\left(\frac{\partial z}{\partial y}\right) = z_{xy} \text{ or } f_{xy}$$

Hence $\frac{\partial^2 z}{\partial y \partial x}$ measures the rate of $\frac{\partial z}{\partial x}$ change of with respect to y while x is held constant; and $\frac{\partial^2 z}{\partial x \partial y}$ measures the rate of $\frac{\partial z}{\partial y}$ change of with respect to x while y is held constant.

Note: If both cross partial derivatives are continuous, they will be identical. i.e., the order of partial derivation is immaterial, provided they are continuous. This result is due to *Young's theorem.*

Example:

If $z = 3x^3 + 5y^3 - x^2y + 2xy + y^2 - 3x - 15$, find

$$\frac{\partial z}{\partial y}, \frac{\partial z}{\partial y}, \frac{\partial^2 z}{\partial x^2}, \frac{\partial^2 z}{\partial y^2}, \frac{\partial^2 z}{\partial y \partial x} \text{ and } \frac{\partial^2 z}{\partial x \partial y}$$

Solution:

$$\frac{\partial z}{\partial y} = 9x^2 - 2xy + 2y - 3$$

$$\frac{\partial z}{\partial y} = 15y^2 - x^2 + 2x + 2y$$

$$\frac{\partial^2 z}{\partial x^2} = \frac{\partial}{\partial x}\left(\frac{\partial z}{\partial x}\right) = 18x - 2y$$

$$\frac{\partial^2 z}{\partial y^2} = \frac{\partial}{\partial y}\left(\frac{\partial z}{\partial y}\right) = 30y + 2$$

$$\frac{\partial^2 z}{\partial y \partial x} = \frac{\partial}{\partial y}\left(\frac{\partial z}{\partial x}\right) = -2x + 2$$

$$\frac{\partial^2 z}{\partial x \partial y} = \frac{\partial}{\partial x}\left(\frac{\partial z}{\partial y}\right) = -2x + 2$$

TOTAL DERIVATIVE

Consider a function: $u = f(x, y)$, where $x = \varphi(t)$ and $y = \psi(t)$. Here if we substitute for x and y in $f(x, y)$, u may be treated as a function of the single variable t. Hence it is possible to find the differential coefficient of u with respect to t, namely $\frac{du}{dt}$. To make it distinct from the partial derivatives of u, we call it as the total differential coefficient (or total derivative) of u.

For finding $\frac{du}{dt}$:

we take Δt as a small increment in t. Correspondingly, let , Δx, Δy and Δu be the changes in x, y and u.

Then,

$$u + \Delta u = f(x + \Delta x, y + \Delta y)$$

$$\therefore \quad \Delta u = f(x + \Delta x, y + \Delta y) - f(x, y)$$

$$= f(x + \Delta x, y + \Delta y) - f(x, y + \Delta y) + f(x, y + \Delta y) - f(x, y),$$

by adding and subtracting the term $f(x, y + \Delta y)$

Hence, $$\frac{\Delta u}{\Delta t} = \frac{f(x+\Delta x, y+\Delta y) - f(x, y+\Delta y)}{\Delta t} + \frac{f(x, y+\Delta y) - f(x, y)}{\Delta t}$$

$$= \frac{f(x+\Delta x, y+\Delta y) - f(x, y+\Delta y)}{\Delta x} \cdot \frac{\Delta x}{\Delta t} + \frac{f(x, y+\Delta y) - f(x, y)}{\Delta y} \cdot \frac{\Delta y}{\Delta t} \quad \text{...(A)}$$

As $\Delta t \to 0$, we get $\Delta x \to 0$, $\Delta y \to 0$ and $\Delta u \to 0$.

Hence, $$\lim_{\Delta t \to 0} \frac{\Delta u}{\Delta t} = \frac{du}{dt},$$

$$\lim_{\Delta t \to 0} \frac{\Delta x}{\Delta t} = \frac{dx}{dt}, \text{ and}$$

$$\lim_{\Delta t \to 0} \frac{\Delta y}{\Delta t} = \lim_{\Delta t \to 0} \frac{\Delta y}{\Delta t}$$

And, $$\lim_{\Delta x \to 0} \frac{f(x+\Delta x, y+\Delta y) - f(x, y+\Delta y)}{\Delta x} = \frac{\partial f(x, y+\Delta y)}{\partial x}$$

But, if we consider $\frac{\partial f(x, y)}{\partial x}$ as a continuous function of y then ,

$$\lim_{\Delta y \to 0} \frac{\partial f(x, y+\Delta y)}{\partial x} = \frac{\partial f(x,y)}{\partial x} = \frac{\partial u}{\partial x} = f_x$$

$$\lim_{\Delta x \to 0} \frac{f(x+\Delta x, y+\Delta y) - f(x, y+\Delta y)}{\Delta x} = f_x$$

Similarly we get,

$$\lim_{\Delta y \to 0} \frac{f(x, y+\Delta y) - f(x,y)}{\Delta y} = \frac{\partial u}{\partial y} = f_y$$

Thus (A) becomes,

$$\frac{du}{dt} = f_x.\frac{dx}{dt} + f_y.\frac{dy}{dt} \qquad ...(1)$$

The above formula is known as the '*Chain Rule*' for partial derivatives. The Chain Rule can be extended for several independent variables.

Suppose $u = f(x_1, x_2, ... x_n)$, where $x_1, x_2, ... x_n$ are all functions of t. Then,

$$\frac{du}{dt} = f_{x_1}.\frac{dx_1}{dt} + f_{x_2}.\frac{dx_2}{dt} + ... + f_{x_n}.\frac{dx_n}{dt}$$

Example 1:

If $u = x^2 + y^2$; where $x = a \cos t$, $y = b \sin t$, find $\frac{du}{dt}$.

Solution:

By Chain Rule,

$$\frac{du}{dt} = f_x.\frac{dx}{dt} + f_y.\frac{dy}{dt}$$

$$u = x^2 + y^2$$

$$f_x = \frac{\partial u}{\partial x} = 2x \; ; f_y = \frac{\partial u}{\partial y} = 2y$$

$$x = a \cos t, \; \frac{dx}{dt}; = -a \sin t$$

$$y = b \sin t, \; \frac{dy}{dt} = b \cos t$$

$$\therefore \quad \frac{du}{dt} = 2x\,(-a \sin t) + 2y.\, b \cos t$$

$$= 2\,a \cos t\,(\text{“}a \sin t) + 2\,b \sin t.\, b \cos t$$

$$= -2\ a^2 \cos t \sin t + 2\ b^2 \sin t.\cos t$$

$$= 2(b^2 - a^2) \sin t.\cos t$$

When $t = x$, the formula (1) becomes:

$$\frac{du}{dx} = f_x + f_y.\frac{dy}{dx} \qquad ...(2)$$

This shows that the rate of change of u with respect to x is the sum of two rates of changes. One is the rate of change of u with respect to x holding y constant and the second is the rate of change of u with respect to y, which in turn is due to the change in x.

Similar result follows for $t = y$, as:

$$\frac{du}{dy} = f_x\frac{dx}{dy} + f_y \qquad ...(3)$$

Consider $u = f(x, y, z)$, where y and z are functions of x. Then the Chain rule for the case $t = x$ may be extended as:

$$\frac{du}{dx} = f_x + f_y.\frac{dy}{dx} + f_z.\frac{dz}{dx} \qquad ...(4)$$

Example 2:

Find $\frac{du}{dx}$ if:

i. $u = x^2 + y^2$ and $y = \sin x$

ii. $u = x^2 + yz$; $y = x^2 + x$, $z = x + 1$

Solution:

i. By Chain Rule,

$$\frac{du}{dx} = f_x + f_y.\frac{dy}{dx}$$

$$u = x^2 + y^2$$

$$f_x = 2x\ ;\ f_y = 2y$$

$$y = \sin x,\ \frac{dy}{dx} = \cos x$$

$$\therefore \quad \frac{du}{dx} = 2x + 2y.\cos x$$

ii. By Chain Rule,

$$\frac{du}{dx} = f_x + f_y.\frac{dy}{dx} + f_z.\frac{dz}{dx}$$

$$u = x^2 + yz$$

$$f_x = 2x \; ; f_y = z \; ; f_z = y$$

$$y = x^2 + x, \quad \frac{dy}{dx} = 2x+1$$

$$z = x+1, \quad \frac{dz}{dx} = 1$$

$$\therefore \qquad \frac{du}{dx} = 2x + z.\,(2x+1) + y.\,1$$

$$= 2x\,(1+z) + (z + y)$$

In the above example, there was only one independent variable, namely t. Let us see the case where there exists more than one independent variable. Let $u = f(x, y)$ is such that $x = \varphi(s,t)$ and $y = \psi(s,t)$, where s and t are two parameters then there exists partial derivatives $\frac{\partial u}{\partial s}$, which is the derivative of u with respect to s holding t constant and $\frac{\partial u}{\partial t}$, which is the derivative of u with respect to t holding s constant. Consequently we get,

$$\frac{\partial u}{\partial s} = f_x.\frac{\partial x}{\partial s} + f_y.\frac{\partial y}{\partial s}; \qquad \text{...(5)}$$

and $$\frac{\partial u}{\partial t} = f_x.\frac{\partial x}{\partial t} + f_y.\frac{\partial y}{\partial t} \qquad \text{...(6)}$$

Example 3:

If $u =$ and $x = t^2 - s^2$; $y = t^2 + s^2$, find $\frac{\partial u}{\partial s}$ and $\frac{\partial u}{\partial t}$

Solution:

$$u = f(x, y) = x^3 + y^3$$

$$f_x = 3x^2 \; ; f_y = 3y^2$$

$$x = t^2 - s^2 \; ; \quad \frac{\partial x}{\partial s} = -2s\,, \quad \frac{\partial x}{\partial t} = 2t$$

$$y = t^2 + s^2 \; ; \quad \frac{\partial y}{\partial s} = 2s\,, \quad \frac{\partial y}{\partial t} = 2t$$

We have,

$$\frac{\partial u}{\partial s} = f_x.\frac{\partial x}{\partial s} + f_y.\frac{\partial y}{\partial s}$$

$$= 3x^2\,(-2s) + 3y^2.\,2s = 6s\,(y^2 - x^2)$$

Also, $$\frac{\partial u}{\partial t} = f_x.\frac{\partial x}{\partial t} + f_y.\frac{\partial y}{\partial t} = 3x^2.\,2t + 3y^2.\,2 = 6t\,(x^2 + y^2)$$

CHAIN RULE FOR IMPLICIT FUNCTIONS

In case we have an implicit relation between x and y of the form: $f(x, y) = c$, where c is a constant, then, for $u = f(x, y) = c$ we have:

$$0 = \frac{\partial u}{\partial x} + \frac{\partial u}{\partial y}.\frac{dy}{dx} = f_x + f_y.\,\frac{dy}{dx}$$

$$\therefore \quad \frac{dy}{dx} = \frac{-f_x}{f_y} \qquad \text{...(7)}$$

PARTIAL DERIVATIVES

Let $z = f(x, y)$ be a continuous function of two independent variables x and y. In this function, if y remains unchanged and x alone varies, then z becomes a continuous function of x alone. If this function possesses a derivative, then this derivative is called the *partial derivative* of z with respect to x; and is denoted as:

$$\frac{\partial z}{\partial x} \quad (\text{or } z_x \text{ or } f_x)$$

Similarly, the partial derivative of z with respect to y is denoted by:

$$\frac{\partial z}{\partial y} \quad (\text{or } z_y \text{ or } f_y)$$

Thus, $$\frac{\partial z}{\partial x} = \lim_{\Delta x \to 0} \frac{f(x+\Delta x, y) - f(x, y)}{\Delta x}, \text{ and}$$

$$\frac{\partial z}{\partial y} = \lim_{\Delta y \to 0} \frac{f(x, y+\Delta y) - f(x, y)}{\Delta y}$$

If u = $f(x, y, z)$, then partial derivative of u with respect to x is denoted as:

$$\frac{\partial u}{\partial x} \quad (\text{or } u_x \text{ or } f_x)$$

Similarly, the partial derivative of u with respect to y is denoted as:

$$\frac{\partial u}{\partial y} \quad (\text{or } u_y \text{ or } f_y)$$

and, the partial derivative of u with respect to z is denoted as:

$$\frac{\partial u}{\partial z} \quad (\text{or } u_z \text{ or } f_z)$$

Example:

Find the partial derivatives of z with respect to x and y if:

$$u = 2x^3 + xy^2 - 4xy + 5x - 1$$

Solution:

$$u = 2x^3 + xy^2 - 4xy + 5x - 1$$

$$\frac{\partial u}{\partial x} = 6x^2 + y^2 - 4y + 5 ;$$

$$\frac{\partial u}{\partial y} = 2xy - 4x$$

Note: If $u = f(x, y)$ is such that it remains unchanged if x and y are interchanged, then $\frac{\partial u}{\partial y}$ is got by interchanging x and y in the expression for $\frac{\partial u}{\partial x}$. In this case $u = f(x, y)$ is called *symmetric* in x and y.

eg: $u = \frac{x^2 + xy + y^2}{x + y}$ is a symmetric function.

PARTIAL DERIVATIVE OF A PRODUCT OR QUOTIENT

As for ordinary differentiation, we can apply the rules of differentiation for partial differentiation also. Product rule and quotient rule of partial differentiation are given below:

If $u(x, y)$ and $v(x, y)$ are two functions of x and y, then,

$$\frac{\partial(uv)}{\partial x} = u.\frac{\partial v}{\partial x} + v.\frac{\partial u}{\partial x} \text{ ; and}$$

$$\frac{\partial(uv)}{\partial y} = u.\frac{\partial v}{\partial y} + v.\frac{\partial u}{\partial y}$$

Also, $$\frac{\partial(u/v)}{\partial x} = \frac{v.\frac{\partial u}{\partial x} - u.\frac{\partial v}{\partial x}}{v^2} \text{ ; and}$$

$$\frac{\partial(u/v)}{\partial y} = \frac{v.\frac{\partial u}{\partial y} - u.\frac{\partial v}{\partial y}}{v^2}$$

Example:

Find f_x and f_y if $f(x, y) = \frac{x^2 + xy + y^2}{x + y}$

Solution:

$$f(x, y) = \frac{x^2 + xy + y^2}{x + y}$$

$$f_x = \frac{(x+y).(2x+y)-(x^2+xy+y^2).1}{(x+y)^2} = \frac{x(x+2y)}{(x+y)^2}$$

$$f_y = \frac{(x+y).(x+2y)-(x^2+xy+y^2).1}{(x+y)^2} = \frac{(2x+y)y}{(x+y)^2}$$

[Note that f_y can also be determined by interchanging x and y in $f_x = \dfrac{x(x+2y)}{(x+y)^2}$]

HOMOGENEOUS FUNCTIONS

A function $z = f(x, y)$ is said to be a homogeneous function of degree n if for $t > 0$, we get the relation: $f(tx, ty) = t^n f(x, y)$.

Example:

Examine whether the following functions are homogeneous.

i. $x^2 + 2xy + y^2$

ii. $2x^2y + xy + 3y^2$

iii. $\dfrac{\sqrt{x^3 + y^3}}{x + y}$

Solution:

i. Let $f(x, y) = x^2 + 2xy + y^2$

Put $x = tx$ and $y = ty$

$$\therefore \quad f(tx, ty) = (tx)^2 + 2(tx)(ty) + (ty)^2$$
$$= t^2 x^2 + 2 t^2 xy + t^2 y^2$$
$$= t^2 (x^2 + 2xy + y^2) = t^2 f(x, y)$$

Hence $f(x, y)$ is homogeneous with degree 2.

ii. Let $f(x, y) = 2x^2y + xy + 3y^2$

Put $x = tx$ and $y = ty$

$$\therefore \quad f(tx, ty) = 2(tx)^2 (ty) + (tx)(ty) + 3(ty)^2$$
$$= 2t^3 x^2 y + t^2 xy + 3t^2 y^2$$
$$\neq t^n f(x, y)$$

Hence $f(x, y)$ is not homogeneous.

iii. Let $f(x, y) = \dfrac{\sqrt{x^3 + y^3}}{x + y}$

Put $x = tx$ and $y = ty$

$$\therefore \quad f(tx, ty) = \frac{\sqrt{(tx)^3 + (ty)^3}}{tx + ty}$$

$$= \frac{\sqrt{t^3x^3 + t^3y^3}}{tx + ty} = \frac{t^{3/2}\sqrt{x^3 + y^3}}{t(x + y)}$$

$$= \frac{t^{1/2}\sqrt{x^3 + y^3}}{(x + y)} \quad t^{1/2} f(x, y)$$

Hence $f(x, y)$ is homogeneous with degree ½.

Result

If $f(x, y)$ is a homogeneous function of degree n, then, it can be expressed in the form:

$$f(x, y) = x^n \varphi\left(\frac{y}{x}\right)$$

For example, in illustration 10.3 we have seen that $f(x, y) = x^2 + 2xy + y^2$ is homogenous of degree 2.

Here $f(x, y)$ can be written as:

$$f(x, y) = x^2 + 2xy + y^2$$

$$= x^2\left[1 + 2\left(\frac{y}{x}\right) + \left(\frac{y}{x}\right)^2\right] = x^2 \varphi\left(\frac{y}{x}\right)$$

SOLVED EXAMPLES

Example 1:

If $u = f(x, y)$ is a homogeneous function of degree zero, show that =

$$\frac{\partial u}{\partial y} = \frac{-x}{y}\frac{\partial u}{\partial x}$$

Solution:

By Euler's Theorem,

$$x.\frac{\partial u}{\partial x} + y.\frac{\partial u}{\partial y} = 0.\ u = 0$$

$$\therefore \qquad y.\frac{\partial u}{\partial y} = -x.\ \frac{\partial u}{\partial x}$$

$$\text{i.e.,} \qquad \frac{\partial u}{\partial y} = \frac{-x}{y}\frac{\partial u}{\partial x}$$

Example 2:

Verify for the following functions, that $f_{yx} = f_{xy}$

i. $f(x, y) = ax^2 \sin y + by^2 \cos x$

ii. $f(x, y) = x^4 + y^4$ iii. $f(x, y) = \dfrac{x^2}{2x+y}$

iv. $f(x, y) = x^y + y^x$ v. $f(x, y) = \dfrac{x}{\sqrt{x}+\sqrt{y}}$

Solution:

i. $f(x, y) = ax^2 \sin y + by^2 \cos x$

$$f_x = \frac{\partial f}{\partial x} = 2ax \sin y - by^2 \sin x$$

$$f_{yx} = \frac{\partial^2 f}{\partial y \partial x} = \frac{\partial}{\partial y}\left(\frac{\partial f}{\partial x}\right) = 2ax \cos y - 2by \sin x \qquad \text{...(A)}$$

$$f_y = \frac{\partial f}{\partial y} = ax^2 \cos y + 2by \cos x$$

$$f_{xy} = \frac{\partial^2 f}{\partial x \partial y} = \frac{\partial}{\partial x}\left(\frac{\partial f}{\partial y}\right) = 2ax \cos y - 2by \sin x \qquad \text{...(B)}$$

Hence $f_{yx} = f_{xy}$

ii. $f(x, y) = x^4 + y^4$

$f_x = 4x^3;\ f_{yx} = 0$ ...(A)

$f_y = 4y^3;\ f_{xy} = 0$ ...(B)

Hence $f_{yx} = f_{xy}$

iii. $f(x, y) = \dfrac{x^2}{2x+y}$

$$f_x = \frac{(2x+y).2x - x^2.2}{(2x+y)^2} = \frac{2x(x+y)}{(2x+y)^2}$$

$$f_{yx} = \frac{(2x+y)^2.2x - 2x(x+y).2(2x+y)}{(2x+y)^4}$$

$$= \frac{-4x^2y - 2xy^2}{(2x+y)^4} = \frac{-2xy}{(2x+y)^3} \quad \ldots(A)$$

$$f_y = \frac{-x^2}{(2x+y)^2}$$

$$f_{xy} = \frac{(2x+y)^2.(-2x) + x^2.2(2x+y).2}{(2x+y)^4}$$

$$= \frac{-4x^2y - 2xy^2}{(2x+y)^4} = \frac{-2xy}{(2x+y)^3} \quad \ldots(B)$$

Hence $f_{yx} = f_{xy}$

iv. $f(x, y) = x^y + y^x$

$$f_x = yx^{y-1} + y^x \log y$$

$$f_{yx} = yx^{y-1}\log x + x^{y-1}.1 + + y^x (1/y) + \log y.\ x\ y^{x-1}$$

$$= x^{y-1}\ y \log x + x^{y-1} + y^{x-1} + x\ y^{x-1} \log y$$

$$= x^{y-1} (y \log x + 1) + y^{x-1} (1 + x \log y) \quad \ldots(A)$$

$$f_y = x^y \log x + x\ y^{x-1}$$

$$f_{xy} = x^y (1/x) + \log x.\ y\ x^{y-1} + x\ y^{x-1} \log y + y^{x-1}$$

$$= x^{y-1} + x^{y-1}\ y \log x + x\ y^{x-1} \log y + y^{x-1}$$

$$= x^{y-1} (1 + y \log x) + y^{x-1} (x \log y + 1) \quad \ldots(B)$$

Hence $f_{yx} = f_{xy}$

v. $f(x, y) = \dfrac{x}{\sqrt{x} + \sqrt{y}}$

$$f_x = \frac{\left(\sqrt{x}+\sqrt{y}\right).1 - x.\dfrac{1}{2\sqrt{x}}}{\left(\sqrt{x}+\sqrt{y}\right)^2} = \frac{\sqrt{x}+2\sqrt{y}}{2\left(\sqrt{x}+\sqrt{y}\right)^2}$$

$$f_y = \frac{2\left(\sqrt{x}+\sqrt{y}\right)^2.\dfrac{1}{\sqrt{y}} - \left(\sqrt{x}+2\sqrt{y}\right).4\left(\sqrt{x}+\sqrt{y}\right)\dfrac{1}{2\sqrt{y}}}{4\left(\sqrt{x}+\sqrt{y}\right)^4}$$

$$= \frac{2\left(\sqrt{x}+\sqrt{y}\right).\dfrac{1}{\sqrt{y}} - \left(\sqrt{x}+2\sqrt{y}\right).\dfrac{2}{\sqrt{y}}}{4\left(\sqrt{x}+\sqrt{y}\right)^3}$$

$$= \frac{\frac{2}{\sqrt{y}}\left(\sqrt{x}+\sqrt{y}-\sqrt{x}-2\sqrt{y}\right)}{4\left(\sqrt{x}+\sqrt{y}\right)^3} = \frac{-1}{2\left(\sqrt{x}+\sqrt{y}\right)^3} \quad \ldots(A)$$

$$f_y = \frac{-x}{\left(\sqrt{x}+\sqrt{y}\right)^2}\cdot\frac{1}{2\sqrt{y}} = \frac{-x}{2\sqrt{y}\left(\sqrt{x}+\sqrt{y}\right)^2}$$

$$f_{xy} = \frac{-1}{2\sqrt{y}}\left[\frac{\left(\sqrt{x}+\sqrt{y}\right)^2 .1 - x.2\left(\sqrt{x}+\sqrt{y}\right).\frac{1}{2\sqrt{x}}}{\left(\sqrt{x}+\sqrt{y}\right)^4}\right]$$

$$= \frac{-1}{2\sqrt{y}}\left[\frac{\sqrt{x}+\sqrt{y}-\sqrt{x}}{\left(\sqrt{x}+\sqrt{y}\right)^3}\right] = \frac{-1}{2\left(\sqrt{x}+\sqrt{y}\right)^3} \quad \ldots(B)$$

Hence $f_{yx} = f_{xy}$

Example 3:

If $u = (x - y) + \log(x + y)$, show that $= \dfrac{\partial^2 u}{\partial x^2} = \dfrac{\partial^2 u}{\partial y^2}$.

Solution:

$$u = (x - y) + \log(x + y)$$

$$\frac{\partial u}{\partial x} = 1 + \frac{1}{x+y}$$

$$\frac{\partial u}{\partial y} = -1 + \frac{1}{x+y}$$

$$\frac{\partial^2 u}{\partial x^2} = \frac{-1}{(x+y)^2} \quad \ldots(A)$$

$$\frac{\partial^2 u}{\partial y^2} = \frac{-1}{(x+y)^2} \quad \ldots(B)$$

Hence $\dfrac{\partial^2 u}{\partial x^2} = \dfrac{\partial^2 u}{\partial y^2}$

Example 4(a):

If $u = \sin(ax + y)$, show that $\dfrac{\partial^2 u}{\partial x^2} = a^2 \dfrac{\partial^2 u}{\partial y^2}$

Solution:

$$u = \sin(ax + y)$$

$$\frac{\partial u}{\partial x} = \cos(ax + y).a$$

$$\frac{\partial u}{\partial y} = \cos(ax + y)$$

$$\frac{\partial^2 u}{\partial x^2} = -a \sin(ax + y).a = -a^2 \sin(ax + y) \quad \text{...(A)}$$

$$\frac{\partial^2 u}{\partial y^2} = -\sin(ax + y) \quad \text{...(B)}$$

Hence $\frac{\partial^2 u}{\partial x^2} = a^2 \frac{\partial^2 u}{\partial y^2}$

Example 4(b):

If $z = \log \sqrt{x^2 + y^2}$, *show that* $\frac{\partial^2 z}{\partial x^2} + \frac{\partial^2 z}{\partial y^2} = 0$

Solution:

$$z = \log \sqrt{x^2 + y^2} = ½ \log(x^2 + y^2)$$

$$\frac{\partial z}{\partial x} = \frac{x}{x^2 + y^2} \qquad \frac{\partial u}{\partial y} = \cos(ax + y)$$

$$\frac{\partial^2 z}{\partial x^2} = -a \sin(ax + y).a = -a^2 \sin(ax + y) \quad \text{...(A)}$$

$$\frac{\partial^2 u}{\partial y^2} = -\sin(ax + y) \quad \text{...(B)}$$

Hence $\frac{\partial^2 u}{\partial x^2} = a^2 \frac{\partial^2 u}{\partial y^2}$

Example 5:

For the following problems find $\frac{du}{dt}$ *using Chain Rule:*

a. $u = x^2 + 2xy + y^2$; *where* $x = t + 1$; $y = t - 1$

b. $u = \sqrt{x^2+y^2}$; where $x = t^3$, $y = t^3 + 3$

c. $u = x^2+y^2$; where $x = e^t \cos t$; $y = e^t \sin t$

d. $u = \dfrac{1}{x^2+y^2}$; where $x = t + 1$, $y = t^2 + 1$

e. $u = x^2+y^2+z^2$; where $x = t$, $y = t^2$, $z = t^3$

Solution:

a. $$u = f(x, y) = x^2+2xy+y^2$$

$$f_x = 2x + 2y \ ; \qquad f_y = 2y + 2x$$

$$x = t + 1, \ \frac{dx}{dt} = 1 \ ; \qquad y = t - 1, \ \frac{dy}{dt} = 1$$

By Chain Rule,

$$\frac{du}{dt} = f_x.\frac{dx}{dt} + f_y.\frac{dy}{dt}$$

$$= (2x + 2y).1 + (2y + 2x).1 = 4(x + y)$$
$$= 4(t + 1 + t - 1) = 8t$$

b. $$u = f(x, y) = \sqrt{x^2+y^2}$$

$$f_x = \frac{x}{\sqrt{x^2+y^2}} \ ; \qquad f_y = \frac{y}{\sqrt{x^2+y^2}} ,$$

$$x = t^3, \ \frac{dx}{dt} = 3t^2 \ ; \qquad y = t^3 + 3, \ \frac{dy}{dt} = 3t^2$$

By Chain Rule,

$$\frac{du}{dt} = f_x.\frac{dx}{dt} + f_y.\frac{dy}{dt}$$

$$= \frac{x}{\sqrt{x^2+y^2}}.\,3t^2 + \frac{y}{\sqrt{x^2+y^2}}.\,3t^2$$

$$= \frac{3t^2(x+y)}{\sqrt{x^2+y^2}} .$$

c. $u = f(x, y) = x^2 + y^2$

$$f_x = 2x \; ; f_y = 2y$$

$$x = e^t \cos t$$

$$\frac{dx}{dt} = e^t(-\sin t) + e^t \cos t = e^t(\cos t - \sin t)$$

$$y = e^t \sin t$$

$$\frac{dy}{dt} = e^t \cos t + e^t \sin t = e^t(\cos t + \sin t)$$

By Chain Rule,

$$\frac{du}{dt} = f_x \cdot \frac{dx}{dt} + f_y \cdot \frac{dy}{dt}$$

$$= 2x \, . \, e^t(\cos t - \sin t) + 2y \, . \, e^t(\cos t + \sin t)$$

$$= 2\, e^t \cos t . e^t(\cos t - \sin t) + 2\, e^t \sin t . e^t(\cos t + \sin t)$$

$$= 2\, e^{2t}(\cos^2 t - \cos t \sin t + \sin t \cos t + \sin^2 t)$$

$$= 2\, e^{2t}$$

d. $u = f(x, y) = \dfrac{1}{x^2 + y^2}$

$$f_x = \frac{-2x}{(x^2+y^2)^2} \; ; \quad f_y = \frac{-2y}{(x^2+y^2)^2}$$

$$x = t + 1, \ \frac{dx}{dt} = 1 \; ; \qquad y = t^2 + 1, \ \frac{dy}{dt} = 2t$$

By Chain Rule,

$$\frac{du}{dt} = f_x \cdot \frac{dx}{dt} + f_y \cdot \frac{dy}{dt}$$

$$= \frac{-2x}{(x^2+y^2)^2} .1 + \frac{-2y}{(x^2+y^2)^2} \, . \, 2t$$

$$= \frac{-2}{(x^2+y^2)^2}(x + 2y\, t)$$

e. $u = x^2 + y^2 + z^2$; where $x = t$, $y = t^2$, $z = t^3$

$$u = f(x, y) = x^2 + y^2 + z^2$$

$$f_x = 2x \; ; f_y = 2y \; ; f_z = 2z$$

$$x = t\ ,\ \frac{dx}{dt} = 1$$

$$y = t^2,\ \frac{dy}{dt} = 2t$$

$$z = t^3\ ,\ \frac{dz}{dt} = 3\,t^2$$

By Chain Rule,

$$\begin{aligned}\frac{du}{dt} &= f_x.\frac{dx}{dt} + f_y.\frac{dy}{dt} + f_z\,.\frac{dz}{dt}\\ &= 2x\,.1 + 2y\,.\,2t + 2z\,.\,3\,t^2\\ &= 2x + 4yt + 6z\,t^2\\ &= 2t + 4\,t^2\,t + 6\,t^3\,t^2\\ &= 2t + 4\,t^3 + 6\,t^5\end{aligned}$$

Example 6:

Find $\frac{du}{dx}$ *if:*

i. $u = ax^2 + bxy + cy^2$; *where* $y = \sqrt{x^2+1}$

ii. $u = x^2 - y^2$; *where* $y = \log x^2$

iii. $u = xy + yz + zx$; $y = 2x+1$, $z = x^2+3$

iv. $u = x^2 + yz$; $y = x^2 + x$; $z = x+1$

Solution:

i. $u = f(x, y) = ax^2 + bxy + cy^2$

$f_x = 2ax + by$; $\quad f_y = bx + 2cy$

$$y = \sqrt{x^2+1},\ \frac{dy}{dx} = \frac{x}{\sqrt{x^2+1}}$$

By Chain Rule,

$$\frac{du}{dx} = f_x + f_y.\frac{dy}{dx} = 2ax + by + (bx + 2cy)\frac{x}{\sqrt{x^2+1}}$$

ii. $u = f(x, y) = x^2 - y^2$

$f_x = 2x$; $f_y = -2y$

$$y = \log x^2 \; ; \; \frac{dy}{dx} = \frac{1}{x^2} \cdot 2x = \frac{2}{x}$$

By Chain Rule,

$$\frac{du}{dx} = f_x + f_y \cdot \frac{dy}{dx}$$

$$= 2x - 2y \cdot \frac{2}{x} = \frac{2}{x}(x^2 - 2y)$$

iii. $u = xy + yz + zx$;

$f_x = y + z$; $f_y = x + z$; $f_z = y + x$

$$y = 2x + 1, \; \frac{dy}{dx} = 2$$

$$z = x^2 + 3, \; \frac{dz}{dx} = 2x$$

By Chain Rule,

$$\frac{du}{dx} = f_x + f_y \cdot \frac{dy}{dx} + f_z \cdot \frac{dz}{dx}$$

$$= (y + z) + (x + z).2 + (y + x).2x$$

$$= 2x(x + y + 1) + (y + 3z)$$

iv. $u = x^2 + yz$; $y = x^2 + x$; $z = x + 1$

$$\frac{du}{dx} = f_x + f_y \cdot \frac{dy}{dx} + f_z \cdot \frac{dz}{dx}$$

$u = x^2 + yz$

$f_x = 2x$; $f_y = z$; $f_z = y$

$$y = x^2 + x, \; \frac{dy}{dx} = 2x + 1$$

$$z = x + 1, \; \frac{dz}{dx} = 1$$

$$\therefore \quad \frac{du}{dx} = 2x + z.(2x+1) + y.1$$

$$= 2x(1+z) + (z + y)$$

Example 7(a):

Find $\frac{\partial u}{\partial s}$ *and* $\frac{\partial u}{\partial t}$ *from the following:*

i. $u =$ *and* $x = t - s$; $y = t + s$

ii. $u = e^{xy}$ *and* $x = t \sin s$; $y = t \cos s$

Solution:

i. $u = f(x, y) = x^2 + xy$

$f_x = 2x + y$; $f_y = x$

$x = t - s$; $\frac{\partial x}{\partial s} = -1$, $\frac{\partial x}{\partial t} = 1$

$y = t + s$; $\frac{\partial y}{\partial s} = 1$, $\frac{\partial y}{\partial t} = 1$

$$\frac{\partial u}{\partial s} = f_x . \frac{\partial x}{\partial s} + f_y . \frac{\partial y}{\partial s}$$

$$= (2x + y)(-1) + x . 1 = -(x + y)$$

$$\frac{\partial u}{\partial t} = f_x . \frac{\partial x}{\partial t} + f_y . \frac{\partial y}{\partial t} = (2x + y) . 1 + x . 1 = 3x + y$$

ii. $u = f(x, y) = e^{xy}$

$f_x = y\, e^{xy}$; $f_y = x\, e^{xy}$

$x = t \sin s$; $\frac{\partial x}{\partial s} = t \cos s$; $\frac{\partial x}{\partial t} = \sin s$

$y = t \cos s$; $\frac{\partial y}{\partial s} = -t \sin s$;

$\frac{\partial y}{\partial t} = \cos s$

$$\frac{\partial u}{\partial s} = f_x . \frac{\partial x}{\partial s} + f_y . \frac{\partial y}{\partial s}$$

$$= y\, e^{xy}\, t \cos s + x\, e^{xy} (-t \sin s)$$

$$= t\, e^{xy} [y \cos s - x \sin s]$$

$$\frac{\partial u}{\partial t} = f_x . \frac{\partial x}{\partial t} + f_y . \frac{\partial y}{\partial t}$$

$$= y\, e^{xy} \sin s + x\, e^{xy} \cos s$$

$$= e^{xy} [y \sin s + x \cos s].$$

Example7(b):

Find the partial derivatives of z with respect to x and y if:

i. $z = x^3 + y^2 - 3xy$

ii. $u = x^3y + y^2 z - 2xz + 5yz - 3xyz - 5z$

Solution:

i. $z = x^3 + y^2 - 3xy$

$$\frac{\partial z}{\partial x} = 3x^2 + 0 - 3y = 3x^2 - 3y \;;$$

$$\frac{\partial z}{\partial y} = 0 + 2y - 3x = 2y - 3x$$

ii. $u = x^3y + y^2 z - 2xz + 5yz - 3xyz - 5z$

$$\frac{\partial u}{\partial x} = 3x^2y + 0 - 2z + 0 - 3yz - 0 = 3x^2y - 2z - 3yz \;;$$

$$\frac{\partial u}{\partial y} = x^3 + 2yz - 0 + 5z - 3xz - 0 = x^3 + 2yz + 5z - 3xz;$$

$$\frac{\partial u}{\partial z} = 0 + y^2 - 2x + 5y - 3xy - 5 = y^2 - 2x + 5y - 3xy - 5$$

Example 8:

Find f_x and f_y if $f(x, y) = (x^2 + 2) \sin xy$

Solution:

$$\begin{aligned} f(x, y) &= (x^2 + 2) \sin xy \\ f_x &= (x^2 + 2). \cos xy. y + \sin xy. 2x \\ &= (x^2 + 2)y. \cos xy + 2x. \sin xy. \\ f_y &= (x^2 + 2). \cos xy. x + \sin xy. 0 \\ &= x(x^2 + 2). \cos xy \end{aligned}$$

Example 9:

If $u = \log \dfrac{x^3 - y^3}{x^2 + y^2}$, *show that* $x\dfrac{\partial u}{\partial x} + y\dfrac{\partial u}{\partial y} = 1$

Solution:

$$\begin{aligned} u &= \log_e \frac{x^3 - y^3}{x^2 + y^2} \\ &= \log_e (x^3 - y^3) - \log_e (x^2 + y^2) \end{aligned}$$

$$\frac{\partial u}{\partial x} = \frac{3x^2}{x^3 - y^3} - \frac{2x}{x^2 + y^2}$$

$$\frac{\partial u}{\partial y} = \frac{-3y^2}{x^3-y^3} - \frac{2y}{x^2+y^2}$$

$$\therefore \quad x\frac{\partial u}{\partial x} + y\frac{\partial u}{\partial x} = x\left[\frac{3x^2}{x^3-y^3} - \frac{2x}{x^2+y^2}\right] + y\left[\frac{-3y^2}{x^3-y^3} - \frac{2y}{x^2+y^2}\right]$$

$$= \frac{3x^3}{x^3-y^3} - \frac{2x^2}{x^2+y^2} - \frac{3y^3}{x^3-y^3} - \frac{2y^2}{x^2+y^2}$$

$$= \frac{3(x^3-y^3)}{x^3-y^3} - \frac{2(x^2+y^2)}{x^2+y^2} = 3 - 2 = 1$$

Example 10:

If $u = \tan^{-1}\frac{x^3+y^3}{x+y}$, *show that* $x\frac{\partial u}{\partial x} + y\frac{\partial u}{\partial x} = \sin 2u$

Solution:

$$u = \tan^{-1}\frac{x^3+y^3}{x+y}$$

$$\Rightarrow \quad \tan u = \frac{x^3+y^3}{x+y} \qquad \ldots(1)$$

Differentiating (1) partially with respect to x,

$$\sec^2 u\frac{\partial u}{\partial x} = \frac{(x+y).3x^2 - (x^3+y^3).1}{(x+y)^2} = \frac{2x^3+3x^2y-y^3}{(x+y)^2}$$

$$x\sec^2 u\frac{\partial u}{\partial x} = x\left[\frac{2x^3+3x^2y-y^3}{(x+y)^2}\right] = \frac{2x^4+3x^3y-xy^3}{(x+y)^2} \qquad \ldots(A)$$

From symmetry,

$$y\sec^2 u\frac{\partial u}{\partial y} = \frac{3xy^3+2y^4-x^3y}{(x+y)^2} \qquad \ldots(B)$$

Adding (A) and (B),

$$x\sec^2 u\frac{\partial u}{\partial x} + y\sec^2 u\frac{\partial u}{\partial y}$$

$$= \frac{2x^4+3x^3y-xy^3}{(x+y)^2} + \frac{3xy^3+2y^4-x^3y}{(x+y)^2}$$

$$= \frac{2x^4 + 3x^3y - xy^3 + 3xy^3 + 2y^4 - x^3y}{(x+y)^2}$$

$$= \frac{2x^4 + 2x^3y + 2xy^3 + 2y^4}{(x+y)^2} = \frac{2x^3(x+y) + 2y^3(x+y)}{(x+y)^2}$$

$$= \frac{2(x+y)(x^3+y^3)}{(x+y)^2} = \frac{2(x^3+y^3)}{(x+y)} = 2 \tan u$$

i.e., $\sec^2 u \left(x\frac{\partial u}{\partial x} + y\frac{\partial u}{\partial y}\right) = 2 \tan u$

$\therefore$ $x\frac{\partial u}{\partial x} + y\frac{\partial u}{\partial y} = \frac{2 \tan u}{\sec^2 u} = 2 \tan u.\cos^2 u$

$$= 2 \sin u. \cos u = \sin 2u$$

Example 11:

Show that the Cobb Douglas production function, $q = AL^a K^{1-a}$ is a homogeneous function of degree 1.

Solution:

$$q = AL^\alpha K^{1-\alpha}$$

put $L = tL$ and $K = tK$

$\therefore$ $q(L, K) = A (tL)^\alpha (tK)^{1-\alpha}$

$$= A t^\alpha L^\alpha t^{1-\alpha} K^{1-\alpha} = t^{\alpha+1-\alpha} AL^\alpha K^{1-\alpha}$$

$$= t^1 AL^\alpha K^{1-\alpha} = t^1 q(L, K)$$

Hence q is homogeneous with degree 1.

Example 12:

Verify Euler's theorem for the following functions:

i. $f(x, y) = ax^2 + 2hxy + by^2$ ii. $f(x, y) = x^3 + 3x^2y + 2xy^2$

iii. $z = x^4 + y^4$ iv. $u(x, y) = \sqrt{ax^2 + 2hxy + by^2}$

v. $f(x, y) = \frac{x^2 + y^2}{2}$ vi. $u(x, y) = \frac{xy}{x+y}$

vii. $z(x, y) = \frac{xy}{x^2+y^2}$ viii. $z(x, y) = \sqrt{\frac{x-y}{x+y}}$

ix. $f(x, y) = \sin\sqrt{\frac{x+y}{x-y}}$ x. $f(x, y) = \frac{x+y}{\sqrt{x}+\sqrt{y}}$

Solution:

i. $f(x, y) = ax^2 + 2hxy + by^2$ ...(A)

Put $x = tx$ and $y = ty$.

Then $f(tx, ty) = a(tx)^2 + 2h(tx)(ty) + b(ty)^2$

$= at^2x^2 + 2ht^2xy + bt^2y^2$

$= t^2 (ax^2 + 2hxy + by^2) = t^2 f(x, y)$

Hence $f(x, y)$ is homogeneous with degree 2.

Now differentiating (A) partially with respect to x,

$$\frac{\partial f}{\partial x} = 2ax + 2hy$$

$$\therefore \quad x\frac{\partial f}{\partial x} = 2ax^2 + 2hxy \quad \text{...(1)}$$

Differentiating (A) partially with respect to y,

$$\frac{\partial f}{\partial y} = 2hx + 2by$$

$$\therefore \quad y\frac{\partial f}{\partial y} = 2hxy + 2by^2 \quad \text{...(2)}$$

$$\text{Hence } x\frac{\partial f}{\partial x} + y\frac{\partial f}{\partial y} = 2ax^2 + 2hxy + 2hxy + 2by^2$$

$$= 2ax^2 + 4hxy + 2by^2$$

$$= 2(ax^2 + 2hxy + by^2) = 2f$$

This verifies Euler's theorem.

ii. $f(x, y) = x^3 + 3x^2y + 2xy^2$ ...(A)

Put $x = tx$ and $y = ty$.

Then $f(tx, ty) = (tx)^3 + 3(tx)^2(ty) + 2(tx)(ty)^2$

$= t^3x^3 + 3t^3x^2y + 2t^3xy^2$

$= t^3 (x^3 + 3x^2y + 2xy^2) = t^3 f(x, y)$

Hence $f(x, y)$ is homogeneous with degree 3.

Now differentiating (A) partially with respect to x,

$$\frac{\partial f}{\partial x} = 3x^2 + 6xy + 2y^2$$

$$\therefore \quad x\frac{\partial f}{\partial x} = 3x^3 + 6x^2y + 2xy^2 \quad \text{...(1)}$$

Differentiating (A) partially with respect to y,

$$\frac{\partial f}{\partial y} = 3x^2 + 4xy$$

$$\therefore \quad y\frac{\partial f}{\partial y} = 3x^2y + 4xy^2 \qquad \ldots(2)$$

Hence $x\dfrac{\partial f}{\partial x} + y\dfrac{\partial f}{\partial y} = 3x^3 + 6x^2y + 2xy^2 + 3x^2y + 4xy^2$

$$= 3x^3 + 9x^2y + 6xy^2$$

$$= 3(x^3 + 3x^2y + 2xy^2) = 3f$$

This verifies Euler's theorem.

iii. $z = x^4 + y^4$ ...(A)

Put x = tx and y = ty

Then $z(\text{tx}, \text{ty}) = (\text{tx})^4 + (\text{ty})^4$

$$= \text{t}^4x^4 + \text{t}^4y^4 = \text{t}^4(x^4 + y^4) = \text{t}^4 z$$

Hence z is homogeneous with degree 4.

Now differentiating (A) partially with respect to x,

$$\frac{\partial z}{\partial x} = 4x^3$$

$$\therefore \quad x\frac{\partial z}{\partial x} = 4x^4 \qquad \ldots(1)$$

From symmetry,

$$y\frac{\partial z}{\partial y} = 4y^4 \qquad \ldots(2)$$

Hence $x\dfrac{\partial z}{\partial x} + y\dfrac{\partial z}{\partial y} = 4x^4 + 4y^4$

$$= 4(x^4 + y^4) = 4z$$

This verifies Euler's theorem.

iv. $u(x, y) = \sqrt{ax^2 + 2hxy + by^2}$ ...(A)

Put x = tx and y = ty.

Then $u(\text{tx}, \text{ty}) = \sqrt{a(tx)^2 + 2h(tx)(ty) + b(ty)^2}$

$$= \sqrt{at^2x^2 + 2htxty + bt^2y^2}$$

$$= t\sqrt{ax^2 + 2hxy + by^2} = t^1\, u(x, y)$$

Hence $u(x, y)$ is homogeneous with degree 1.

Now differentiating (A) partially with respect to x,

$$\frac{\partial u}{\partial x} = \frac{2ax + 2hy}{2\sqrt{ax^2 + 2hxy + by^2}} = \frac{ax + hy}{\sqrt{ax^2 + 2hxy + by^2}}$$

$$\therefore \quad x\frac{\partial u}{\partial x} = 3x^3 + 6x^2y + 2xy^2 \quad \text{...(1)}$$

Differentiating (A) partially with respect to y,

$$\frac{\partial u}{\partial y} = 3x^2 + 4xy$$

$$\therefore \quad y\frac{\partial u}{\partial y} = 3x^2y + 4xy^2 \quad \text{...(2)}$$

$$\text{Hence } x\frac{\partial u}{\partial x} + y\frac{\partial u}{\partial y} = 3x^3 + 6x^2y + 2xy^2 + 3x^2y + 4xy^2$$

$$= 3x^3 + 9x^2y + 6xy^2$$

$$= 3(x^3 + 3x^2y + 2xy^2) = 3u$$

This verifies Euler's theorem.

$$\text{v. } f(x, y) = \frac{x^2 + y^2}{2} \quad \text{...(A)}$$

Put $x = tx$ and $y = ty$.

$$\text{Then} \quad f(tx, ty) = \frac{(tx)^2 + (ty)^2}{2}$$

$$= \frac{t^2x^2 + t^2y^2}{2} = \frac{t^2(x^2 + y^2)}{2} = t^2 f(x, y)$$

Hence $f(x, y)$ is homogeneous with degree 2.

Now differentiating (A) partially with respect to x,

$$\frac{\partial f}{\partial x} = \frac{2x}{2} = x$$

$$\therefore \quad x\frac{\partial f}{\partial x} = x^2 \quad \text{...(1)}$$

By symmetry,

$$y\frac{\partial f}{\partial y} = y^2 \qquad ...(2)$$

Hence $x\dfrac{\partial f}{\partial x} + y\dfrac{\partial f}{\partial y} = x^2 + y^2 = 2\,f(x, y)$

This verifies Euler's theorem.

vi. $u(x, y) = \dfrac{xy}{x+y}$ ...(A)

Put $x = tx$ and $y = ty$.

Then $\quad u(tx, ty) = \dfrac{txty}{tx+ty}$

$$= \frac{t^2xy}{t(x+y)} = \frac{txy}{x+y} = t^1\,u(x, y)$$

Hence $u(x, y)$ is homogeneous with degree 1.

Now differentiating (A) partially with respect to x,

$$\frac{\partial u}{\partial x} = \frac{(x+y)y - xy}{(x+y)^2} = \frac{y^2}{(x+y)^2}$$

$$\therefore \qquad x\frac{\partial u}{\partial x} = \frac{xy^2}{(x+y)^2} \qquad ...(1)$$

By symmetry,

$$y\frac{\partial u}{\partial y} = \frac{x^2y}{(x+y)^2} \qquad ...(2)$$

Hence $x\dfrac{\partial u}{\partial x} + y\dfrac{\partial u}{\partial y} = \dfrac{xy^2}{(x+y)^2} + \dfrac{x^2y}{(x+y)^2}$

$$= \frac{xy^2 + x^2y}{(x+y)^2} = \frac{xy(x+y)}{(x+y)^2} = \frac{xy}{x+y} = 2\,u(x, y)$$

This verifies Euler's theorem.

vii. $z(x, y) = \dfrac{xy}{x^2+y^2}$ ...(A)

Put $x = tx$ and $y = ty$.

Then $\quad z(tx, ty) = \dfrac{txty}{(tx)^2 + (ty)^2}$

$$= \frac{t^2 xy}{t^2(x^2+y^2)} = t^0 \frac{xy}{x^2+y^2}$$

$$= t^0 z(x, y)$$

Hence z is homogeneous with degree 0.

Now differentiating (A) partially with respect to x,

$$\frac{\partial z}{\partial x} = \frac{(x^2+y^2)y - xy.2x}{(x^2+y^2)^2}$$

$$= \frac{x^2y + y^3 - 2x^2y}{(x^2+y^2)^2}. = \frac{y^3 - x^2y}{(x^2+y^2)^2}$$

$$\therefore \quad x\frac{\partial z}{\partial x} = \frac{xy^3 - x^3y}{(x^2+y^2)^2} \quad \text{...(1)}$$

By symmetry,

$$y\frac{\partial z}{\partial y} = \frac{x^3y - xy^3}{(x^2+y^2)^2} \quad \text{...(2)}$$

$$\text{Hence } x\frac{\partial z}{\partial x} + y\frac{\partial z}{\partial y} = \frac{xy^3 - x^3y}{(x^2+y^2)^2} + \frac{x^3y - xy^3}{(x^2+y^2)^2}$$

$$= \frac{xy^3 - x^3y + x^3y - xy^3}{(x^2+y^2)^2}$$

$$= 0 = 0.\ z(x, y)$$

This verifies Euler's theorem.

viii. $z(x, y) = \sqrt{\dfrac{x-y}{x+y}}$ ...(A)

Put $x = tx$ and $y = ty$.

$$\text{Then} \quad z(tx, ty) = \sqrt{\frac{tx-ty}{tx+ty}}$$

$$= \sqrt{\frac{t(x-y)}{t(x+y)}} = \sqrt{\frac{x-y}{x+y}}$$

$$= t^0 z(x, y)$$

Hence z is homogeneous with degree 0.

Here $z(x, y) = \sqrt{\dfrac{x-y}{x+y}} = \dfrac{\sqrt{x-y}}{\sqrt{x+y}}$

Now differentiating partially with respect to x,

$$= \frac{\sqrt{x+y}\dfrac{1}{2\sqrt{x-y}} - \sqrt{x-y}\dfrac{1}{2\sqrt{x+y}}}{x+y}$$

$$= \frac{\dfrac{(x+y)-(x-y)}{2\sqrt{(x-y)(x+y)}}}{x+y} = \frac{x+y-x+y}{2(x+y)\sqrt{x^2-y^2}}$$

$$= \frac{2y}{2(x+y)\sqrt{x^2-y^2}} = \frac{y}{(x+y)\sqrt{x^2-y^2}}$$

$$\therefore \quad x\frac{\partial z}{\partial x} = \frac{xy}{(x+y)\sqrt{x^2-y^2}} \qquad \text{...(1)}$$

Differentiating partially with respect to y,

$$\frac{\partial z}{\partial y} = \frac{\sqrt{x+y}\dfrac{-1}{2\sqrt{x-y}} - \sqrt{x-y}\dfrac{1}{2\sqrt{x+y}}}{x+y}$$

$$= \frac{-x-y-x+y}{2(x+y)\sqrt{x^2-y^2}} = \frac{-x}{(x+y)\sqrt{x^2-y^2}}$$

$$\therefore \quad y\frac{\partial z}{\partial y} = \frac{-xy}{(x+y)\sqrt{x^2-y^2}} \qquad \text{...(2)}$$

$$\text{Hence } x\frac{\partial z}{\partial x} + y\frac{\partial z}{\partial y} = \frac{xy}{(x+y)\sqrt{x^2-y^2}} + \frac{-xy}{(x+y)\sqrt{x^2-y^2}}$$

$$= 0 = 0.\ z(x, y)$$

This verifies Euler's theorem.

ix. $f(x, y) = \sin\sqrt{\dfrac{x+y}{x-y}}$ ...(A)

Put $x = tx$ and $y = ty$.

Then, $f(tx, ty) = \sin\sqrt{\dfrac{tx+ty}{tx-ty}} = \sin\sqrt{\dfrac{x+y}{x-y}} = t^0 f(x, y)$

Hence $f(x, y)$ is homogeneous with degree 0.

Here $f(x, y) = \sin\sqrt{\dfrac{x-y}{x+y}} = \sin\sqrt{\dfrac{x-y}{x+y}}$

Now differentiating partially with respect to x,

$$\frac{\partial f}{\partial x} = \cos\frac{\sqrt{x-y}}{\sqrt{x+y}} \times \frac{\sqrt{x-y}\dfrac{1}{2\sqrt{x+y}} - \sqrt{x+y}\dfrac{1}{2\sqrt{x-y}}}{x-y}$$

$$= \cos\frac{\sqrt{x-y}}{\sqrt{x+y}} \times \frac{-y}{(x-y)\sqrt{x^2-y^2}}$$

$$\therefore \quad x\frac{\partial f}{\partial x} = \cos\frac{\sqrt{x-y}}{\sqrt{x+y}} \times \frac{-xy}{(x-y)\sqrt{x^2-y^2}} \quad \text{...(1)}$$

Differentiating partially with respect to y,

$$\frac{\partial f}{\partial y} = \cos\frac{\sqrt{x-y}}{\sqrt{x+y}} \times \frac{\sqrt{x-y}\dfrac{1}{2\sqrt{x+y}} - \sqrt{x+y}\dfrac{-1}{2\sqrt{x-y}}}{x-y}$$

$$= \cos\frac{\sqrt{x-y}}{\sqrt{x+y}} \times \frac{x}{(x-y)\sqrt{x^2-y^2}}$$

$$\therefore \quad y\frac{\partial f}{\partial y} = \cos\frac{\sqrt{x-y}}{\sqrt{x+y}} \times \frac{xy}{(x-y)\sqrt{x^2-y^2}} \quad \text{...(2)}$$

Hence, $x\dfrac{\partial f}{\partial x} + y\dfrac{\partial f}{\partial y}$

$$= \cos\frac{\sqrt{x-y}}{\sqrt{x+y}} \times \frac{-xy}{(x-y)\sqrt{x^2-y^2}} + \cos\frac{\sqrt{x-y}}{\sqrt{x+y}} \times \frac{xy}{(x-y)\sqrt{x^2-y^2}}$$

$= 0 = 0.\ f(x, y)$

This verifies Euler's theorem.

x. $f(x, y) = \dfrac{x+y}{\sqrt{x}+\sqrt{y}}$ ...(A)

Put $x = tx$ and $y = ty$.

Then, $f(tx, ty) = \dfrac{tx+ty}{\sqrt{tx}+\sqrt{ty}}$

$$= \frac{t(x+y)}{\sqrt{t}\left(\sqrt{x}+\sqrt{y}\right)} = \frac{\sqrt{t}(x+y)}{\left(\sqrt{x}+\sqrt{y}\right)} = t^{1/2}f(x, y)$$

Hence $f(x, y)$ is homogeneous with degree 1/2.

Now differentiating (A) partially with respect to x,

$$\frac{\partial f}{\partial x} = \frac{\left(\sqrt{x}+\sqrt{y}\right)-(x+y)\dfrac{1}{2\sqrt{x}}}{\left(\sqrt{x}+\sqrt{y}\right)^2}$$

$$= \frac{2\sqrt{x}\left(\sqrt{x}+\sqrt{y}\right)-(x+y)}{2\sqrt{x}\left(\sqrt{x}+\sqrt{y}\right)^2}$$

$$= \frac{2x+2\sqrt{xy}-x-y}{2\sqrt{x}\left(\sqrt{x}+\sqrt{y}\right)^2} = \frac{x-y+2\sqrt{xy}}{2\sqrt{x}\left(\sqrt{x}+\sqrt{y}\right)^2}$$

$$\therefore \quad x\frac{\partial f}{\partial x} = \frac{x\left(x-y+2\sqrt{xy}\right)}{2\sqrt{x}\left(\sqrt{x}+\sqrt{y}\right)^2} = \frac{\sqrt{x}\left(x-y+2\sqrt{xy}\right)}{2\left(\sqrt{x}+\sqrt{y}\right)^2} \qquad \ldots(1)$$

By symmetry,

$$y\frac{\partial f}{\partial x} = \frac{\sqrt{y}\left(y-x+2\sqrt{xy}\right)}{2\left(\sqrt{x}+\sqrt{y}\right)^2} \qquad \ldots(2)$$

Hence,

$$x\frac{\partial f}{\partial x} + y\frac{\partial f}{\partial x} = \frac{\sqrt{x}\left(x-y+2\sqrt{xy}\right)}{2\left(\sqrt{x}+\sqrt{y}\right)^2} + \frac{\sqrt{y}\left(y-x+2\sqrt{xy}\right)}{2\left(\sqrt{x}+\sqrt{y}\right)^2}$$

$$= \frac{\sqrt{x}\left(x-y+2\sqrt{xy}\right)+\sqrt{y}\left(y-x+2\sqrt{xy}\right)}{2\left(\sqrt{x}+\sqrt{y}\right)^2}$$

$$= \frac{x\sqrt{x}-y\sqrt{x}+2x\sqrt{y}+y\sqrt{y}-x\sqrt{y}+2y\sqrt{x}}{2\left(\sqrt{x}+\sqrt{y}\right)^2}$$

$$= \frac{x\sqrt{x} + x\sqrt{y} + y\sqrt{y} + y\sqrt{x}}{2\left(\sqrt{x}+\sqrt{y}\right)^2}$$

$$= \frac{(x+y)\left(\sqrt{x}+\sqrt{y}\right)}{2\left(\sqrt{x}+\sqrt{y}\right)^2} = \frac{x+y}{2\left(\sqrt{x}+\sqrt{y}\right)} = ½\, f(x, y)$$

This verifies Euler's theorem.

EXERCISES

1. If u = log (tan x + tan y) , show that sin $2x\frac{\partial u}{\partial x}$ + sin $2y\frac{\partial u}{\partial y}$ = 2
2. If u = log (tan x + tan y + tan z),

   show that sin $2x\frac{\partial u}{\partial x}$ + sin $2y\frac{\partial u}{\partial y}$ + sin $2z\frac{\partial u}{\partial z}$ = 2
3. If V = log $\frac{x^2+y^2}{x+y}$, show that $x\frac{\partial V}{\partial x} + y\frac{\partial V}{\partial y} = 1$
4. If V = log ($x^3 + y^3 + z^3 - 3xyz$), show that $\frac{\partial V}{\partial y}+\frac{\partial V}{\partial y}+\frac{\partial V}{\partial z}=\frac{3}{x+y+z}$
5. If U = ($y - z$) ($z - x$) ($x - y$) , show that $\frac{\partial U}{\partial x}+\frac{\partial U}{\partial y}+\frac{\partial U}{\partial z}=2$
6. Use Euler's Theorem on homogeneous functions to prove that,

   $x\frac{\partial u}{\partial x} + y\frac{\partial u}{\partial y}$ = sin 2u, if u = $\tan^{-1}\frac{x^3+y^3}{x+y}$
7. If z = sin (x + y) + log (x + y), show that $\frac{\partial^2 z}{\partial x^2}=\frac{\partial^2 z}{\partial y^2}$
8. If $u = e^{xy}$, show that $\frac{\partial^2 u}{\partial x^2}+\frac{\partial^2 u}{\partial y^2}=\frac{1}{u}\left[\left(\frac{\partial u}{\partial x}\right)^2+\left(\frac{\partial u}{\partial y}\right)^2\right]$
9. If $z = e^x(x\cos y - y\sin y)$, show that $\frac{\partial^2 z}{\partial x^2}+\frac{\partial^2 z}{\partial y^2}=0$
10. If $u = \tan^{-1}(y/x)$, show that $\frac{\partial^2 u}{\partial x^2}+\frac{\partial^2 u}{\partial y^2}=0$

11. If $u = \log(x^2 + y^2 + z^2)$, show that $\frac{\partial^2 u}{\partial x^2} + \frac{\partial^2 u}{\partial x^2} + \frac{\partial^2 u}{\partial z^2} = \frac{2}{x^2 + y^2 + z^2}$

12. If $V = \frac{1}{\sqrt{x^2 + y^2 + z^2}}$, show that $\frac{\partial^2 V}{\partial x^2} + \frac{\partial^2 V}{\partial y^2} + \frac{\partial^2 V}{\partial z^2} = 0$

13. If $u = \sin^{-1}\frac{x}{y} + \tan^{-1}\frac{y}{x}$, show that $x\frac{\partial u}{\partial x} + y\frac{\partial u}{\partial y} = 0$

14. If $u = \sin^{-1}(x - y)$, show that $x\frac{\partial u}{\partial x} + y\frac{\partial u}{\partial y} = \tan u$

15. If $u = \tan^{-1}\frac{x^3 + y^3}{x - y}$, show that $x\frac{\partial u}{\partial x} + y\frac{\partial u}{\partial y} = \sin 2u$

16. If $u = \sin^{-1}\frac{x^2 + y^2}{x + y}$, show that $x\frac{\partial u}{\partial x} + y\frac{\partial u}{\partial y} = \tan u$

5

Functions of Two or More Independent Variable

INTRODUCTION

Let $y = f_1(x)/f_2(x)$. Then $y + \delta y = f_1(x + \delta x)/ f_2(x + \delta x)$.

$$\therefore \qquad = \lim_{\delta x \to 0}\left[\frac{f_1(x+\delta x).f_2(x)-f_1(x).f_2(x+\delta x)}{\delta x.f_2(x+\delta x).f_2(x)}\right]$$

$$\frac{d}{dx}\left[\frac{f_1(x)}{f_2(x)}\right] = \frac{f_2(x).\frac{d}{dx}.f_1(x)-f_1(x).\frac{d}{dx}f_2(x)}{[f_2(x)]^2}.$$

This can also be written as.

$\therefore$ the diff. coeff. of the quotient of two functions

$$= \frac{\text{(diff. coeff. of Nr.) (Dr.) – (Nr.) (diff. coeff. of Dr.)}}{\text{square of the denominator}}$$

DIFFERENTIAL COEFFICIENT OF THE PRODUCT OF A CONSTANT AND A FUNCTION

We have,

$$\frac{d}{dx}\{cf(x)\} = \lim_{\delta x \to 0}\frac{cf(x+\delta x)-cf(x)}{\delta x}$$

$$c\lim_{\delta x \to 0}\frac{f(x+\delta x)-f(x)}{\delta x}\ c\frac{d}{dx}f(x)$$

$$\frac{d}{dx}[cf(x)] = c - \frac{d}{dx}[f(x)].$$

FUNCTIONS OF TWO OR MORE INDEPENDENT VARIABLES

We after come across which depend two or more variables. For example, volume of a right circular cylinder of radius r and height h is given by $v = \pi r^2 h$ for a given value of r and h V has a definite values.

A function of x, y and z may be written as f(x, y, z) or ϕ(x, y, z), etc. Similarly, a function of x and y is generally denoted by the symbols f(x, y) or ϕ(x, y) etc.

If a derivative of a function of several independent variables be found with respect to any of them, keeping the others as constants, it is said to be a Partial Derivative.

The operation of finding the partial derivatives of a function of more than one independent variables is called *Partial Differential.* The symbols $\partial/\partial x$, $\partial/\partial y$, etc., areused to denote such differentiations and the represents $\partial u/\partial x$, $\partial u/\partial y$, etc., are respectively called the *Partial Differential Coefficients* of u w.r.t. x, y, etc. Thus, if u = f(x, y, z), then the partial differential coefficient of u w.r.t. x *i.e.* $\partial u/\partial x$ is obtained by differentiating u w.r.t. x keeping y and z as constants. Sometimes $\partial f/\partial x$ is also denoted by f_x.

Second Order Partial Differential Coefficiens: If u = f(x, y), then $\partial u/\partial x$ or f_x and $\partial u/\partial y$ or f_y are themselves functions of x and y and can be again differentiated partially.

We call $\frac{\partial}{\partial x}\left(\frac{\partial u}{\partial y}\right), \frac{\partial}{\partial y}\left(\frac{\partial u}{\partial x}\right), \frac{\partial}{\partial x}\left(\frac{\partial u}{\partial x}\right)$ and $\frac{\partial}{\partial y}\left(\frac{\partial u}{\partial y}\right)$

as the second order partial derivatives are u and these are respectively as denoted by $\frac{\partial^2 u}{\partial x\, \partial y}, \frac{\partial^2 u}{\partial y\, \partial x}, \frac{\partial^2 u}{\partial x^2}$ and $\frac{\partial^2 u}{\partial y^2}$

Note: If u = f(x, y) and its partial derivatives are continuous (as is true in all ordinary cases), the order of differentiation is immaterial *i.e.*, $\frac{\partial^2 u}{\partial x\, \partial y} = \frac{\partial^2 u}{\partial y\, \partial x}$. The partial differentiation co-efficient of fx and fy are, fxu, fxy, fyx, fyy or

$\frac{\partial^2 f}{\partial x^2}, \frac{\partial^2 f}{\partial y\, \partial x}, \frac{\partial^2 f}{\partial x\, \partial y}, \frac{\partial^2 f}{\partial y^2}$ respectively. It should be specially noted that

$$\frac{\partial^2 f}{\partial y\, \partial x} \text{ means } \frac{\partial}{\partial y}\left(\frac{\partial f}{\partial x}\right) \text{ and } \frac{\partial^2 f}{\partial x\, \partial y} \text{ means } \frac{\partial}{\partial x}\left(\frac{\partial f}{\partial y}\right).$$

The student will be able to convince himself that in all ordinary cases.

$$\frac{\partial^2 f}{\partial y\, \partial x} = \frac{\partial^2 f}{\partial x\, \partial y}.$$

DIFFERENTIAL OF A SUM OR A DIFFERENCE OF TWO FUNCTIONS

Let $\quad f(x) = f_1(x) \pm f_2(x)$.

Then $f(x + \delta x) = f_1(x + \delta x) \pm f_2(x + \delta x)$.

$$\therefore \qquad \frac{dy}{dx} f(x) = \lim_{\delta x \to 0} \frac{\{f_1(x+\delta x) \pm f_2(x+\delta x) - f_1(x) \pm f_2(x)\}}{\delta x},$$

$$= \frac{d}{dx} f_1(x) \pm \frac{d}{dx} f_2(x)$$

$$\Rightarrow \quad \frac{d}{dx}[f_1(x) \pm f_2(x)] = \frac{d}{dx}.[f_1(x)] + \frac{d}{dx}[f_2(x)].$$

DIFFRENTIATION

If f is a function of x

$\lim_{h \to 0} \dfrac{f(a+h) - f(a)}{h}$ is called the differential coefficient of f (x) at x = a. It is generally denoted by f '(a).

The differential coefficient is also called the derivative, or the derived function. We shall use the latter term only for the set of ordered pairs (x, f '(x) and denoted by f '.

The process of finding the differential coefficient is called differentiation. We are said to differential f (x).

It should be carefully noted that in taking the limit {f (a + h) – f (a)}/h is regarded as a function of h, h is regarded as a variable; and a is regarded as a constant. The sequal will show that $\lim_{h \to 0}$ [f (a + h) – f (a)] h is independent of h.

Generally it is more convenient to write x at self for a. It being understood that in the process of finding the limit when h → 0, x is to be kept constant with this understanding we can write :

The differential coefficient of f (x) $= \lim_{h \to 0} \dfrac{f(x+h) - f(x)}{h}$

The differential coefficient of f (x) is written as $\dfrac{df(x)}{dx}$ or $\dfrac{d}{dx}$[f (x)] or f '(x) or D[f(x)].

The differential coefficient of x = a is generally reperesented by f '(x)

or $\left[\dfrac{df(x)}{dx}\right]x = a.$

HOMOGENEOUS FUNCTIONS

An expression in which every term is of the same degree is called a homogeneous function. Thus

$$a_0x^n + a_1x^{n-1}y + a_2x^{n-2}y^2 + ... + a_{n-1}xy^{n-1} + a_ny^n$$

is a homogeneous function of x and y of degree n. This can also be written as

$$x^n\left\{a_0 + a_1\left(\frac{y}{x}\right) + a_2\left(\frac{y}{x}\right)^2 +...+ a_{n-1}\left(\frac{y}{x}\right)^{n-1} + a_n\left(\frac{y}{x}\right)^n\right\}$$

or $x^n f\left(\dfrac{y}{x}\right)$, where $f\left(\dfrac{y}{x}\right)$ is some function of $\left(\dfrac{y}{x}\right)$.

Note 1: *To test whether a given function f(x, y) is homogeneous or not we put tx for x and ty for y in it.*

If we get $f(tx, ty) = t^n f(x, y)$,

the function f(x, y) is homogeneous of degree n; otherwise f(x, y) is not a homogeneous function.

Note 2: *If u is a homogeneous function of x and y of degee n then* $\dfrac{\partial u}{\partial x}$ *and* $\dfrac{\partial u}{\partial y}$ *are also homogeneous functions of x and y each being of degree n – 1.*

Let $u = x^n f\left(\dfrac{y}{x}\right)$.

[∵ u is a homogeneous function of x and y of degree n]

Then $\dfrac{\partial u}{\partial x} = nx^{n-1} f\left(\dfrac{y}{x}\right) + x^n\left\{f'\left(\dfrac{y}{x}\right)\right\}.\left(-\dfrac{y}{x^2}\right)$

$$= x^{n-1}\left[n\, f\left(\frac{y}{x}\right) - \left(\frac{y}{x}\right) f'\left(\frac{y}{x}\right)\right]$$

$= x^{n-1}.\left[\text{some function of } \dfrac{y}{x}\right]$

= a homogeneous function of x and y of degree (n – 1).

Similarly, $\dfrac{\partial u}{\partial y}=x^n\left\{f'\left(\dfrac{y}{x}\right)\right\}.\dfrac{1}{x}=x^{n-1}f'\left(\dfrac{y}{x}\right)$

$= x^{n-1}.\left(\text{some function of } \dfrac{y}{x}\right)$

= a homogeneous function of x and y of degree (n – 1).

EULER'S THEOREM ON HOMOGENEOUS FUNCTIONS

If u is a homogeneous function of x and y of degree n, then

$$x\frac{\partial u}{\partial x}+y\frac{\partial u}{\partial y}=nu.$$

Proof:

Since u is a homogeneous function of x and y of degree n, therefore u may be put in the form

$$u=x^n f\left(\frac{y}{x}\right). \qquad ...(1)$$

Differentiating (1) partially w.r.t. 'x', we hae

$$\frac{\partial u}{\partial x}=\frac{\partial}{\partial x}\left[x^n f\left(\frac{y}{x}\right)\right]$$

$$=\left[f\left(\frac{y}{x}\right)\right]nx^{n-1}+x^n\left[f'\left(\frac{y}{x}\right)\right]\left(-\frac{y}{x^2}\right).$$

$$\therefore\ x\frac{\partial u}{\partial x}=nx^n f\left(\frac{y}{x}\right)-x^{n-1}y.f'\left(\frac{y}{x}\right). \qquad ...(2)$$

Again differentiating (1) partially w.r.t. 'y', we have

$$\frac{\partial u}{\partial y}=\frac{\partial}{\partial y}\left[x^n f\left(\frac{y}{x}\right)\right]=x^n\left[f'\left(\frac{y}{x}\right)\right].\frac{1}{x}=x^{n-1}f'\left(\frac{y}{x}\right).$$

$$\therefore\ y\frac{\partial u}{\partial y}=y.x^{n-1}f'\left(\frac{y}{x}\right). \qquad ...(3)$$

Adding (2) and (3), we have

$= 2xy\ u^3 + 3y^2u^5\ [(x - y)^2 - (1 - 2xy + y^2)],$

$[\because u^{-2} = 1 - 2xy + y^2]$

$= 2xy\ u^3 + 3y^2u^5\ [x^2 - 1] = 2xy\ u^3 - 3y^2u^5\ (1 - x^2).$ ...(2)

Adding (1) and (2), we have

$$\frac{\partial}{\partial x}\left\{\left(1-x^2\right)\frac{\partial u}{\partial x}\right\}+\frac{\partial}{\partial y}\left\{y^2\frac{\partial u}{\partial y}\right\}=0.$$

Example 1:

If $\theta = t^n e^{-r^2-4t}$, *what value of n will make*

$$\frac{1}{r^2}\frac{\partial}{\partial r}\left(r^2\frac{\partial\theta}{\partial r}\right)=\frac{\partial\theta}{\partial t}\,?$$

Solution:

We have $\dfrac{\partial\theta}{\partial r}=t^n.e^{-r^2/4}.\left(-\dfrac{2r}{4t}\right)=-\dfrac{r}{2}t^{n-1}e^{-r^2/4t}.$

$\therefore \quad r^2\dfrac{\partial\theta}{\partial r}=-\dfrac{1}{2}r^3\,t^{n-1}\,e^{-r^2/4t}.$

$\therefore \quad \dfrac{\partial}{\partial r}\left(r^2\dfrac{\partial\theta}{\partial r}\right)=-\dfrac{3r^2}{2}t^{n-1}e^{-r^2/4t}-\dfrac{1}{2}r^3t^{n-1}\,e^{-r^2/4t}.\left(-\dfrac{2r}{4t}\right)$

$=-\dfrac{3}{2}r^2t^{n-1}\,e^{-r^2/4t}+\dfrac{1}{4}r^4t^{n-2}\,e^{-r^2/4t}.$

$\therefore \quad \dfrac{1}{r^2}\dfrac{\partial}{\partial r}\left(r^2\,\dfrac{\partial\theta}{\partial r}\right)=-\dfrac{3}{2}t^{n-1}e^{-r^2/4t}+\dfrac{1}{2}r^2t^{n-2}\,e^{-r^2/4t}$

Also $\quad \dfrac{\partial\theta}{\partial r}\;nt^{n-1}e^{-r^2/4t}+t^n\,e^{-r^2/4t}\;.\dfrac{r^2}{4t^2}$

$= n\;t^{n-1}\;e^{-r2}+1/4\;r^2\,t^{n-2}\;e^{-r2/4t}.$

Now $\quad \dfrac{1}{r^2}\dfrac{\partial}{\partial r}\left(r^2\,\dfrac{\partial\theta}{\partial r}\right)=\dfrac{\partial\theta}{\partial t}$

$\Rightarrow -\dfrac{3}{2}t^{n-1}e^{-r^2/4t}+\dfrac{1}{4}r^2t^{n-2}e^{-r^2/4t}=nt^{n-1}\,e^{-r^2/4t}\;+\dfrac{1}{4}r^2\,t^{n-2}\,e^{-r^2/4t}$

$\Rightarrow -\dfrac{3}{2}t^{n-1}\,e^{-r^2/4t}=nt^{n-1}\,e^{-r^2/4t}$, for all posible values of r and t

$\Rightarrow n=-\dfrac{3}{2}.$

INVERSE HYPERBOLIC FUNCTIONS

Let $y = \sin h^{-1}\,x$. Then $\sin h\;y = x$. Therefore

$\cosh y = \sqrt{(\sin h^2\,y + 1)} = \sqrt{(x^2 + 1)}$. Adding these, we get $\sin h\;y + \cosh y = x + \sqrt{(x^2 + 1)}$

or$1/2(e^y - e^{-y}) + 1/2(e^y - e^{-y}) = x + (x^2 + 1)$

or$e^y = x + \sqrt{(x^2 + 1)}$.

$\therefore y = \log\,[x + \sqrt{(x^2 + 1)}]$

Thus $\sin h^{-1} x = \log [x + \sqrt{(x^2 + 1)}]$

Similarly, $\cos h^{-1} x = \log [x + \sqrt{(x^2 - 1)}]$;

$$\tan h^{-1} x = \frac{1}{2} \log \frac{1+x}{1-x};\ \coth^{-1} x = \frac{1}{2} \log \frac{x+1}{x-1};$$

$$\sec h^{-1} x = \log \frac{1+\sqrt{(1-x^2)}}{x};\ \text{cosech}^{-1} x = \log \frac{1+\sqrt{(1+x^2)}}{x}.$$

SOME MORE METHODS OF DIFFERENTIATION

(a) *Logarithmic Differentiation :* When a function consists of the product or the quotient of a number of functions, we take the logarithm and differentiate. This process, called *logarithmic differentiation,* is also useful when a function of x is raised to a power which is itself a function of x.

(b) *Implict Function:* If the relation between y and x is represented by an equation from which y cannot be easily expressed in terms of x, then y is called an *implicit function* of x. But if x is given directly in terms of x, then y is called an *explicit function* of x.

To find dy/dx in the case of implicit functions, differentiate each term of the given equation w.r.t. 'x' and solve the resulting equation for dy/dx.

Note: Here the value of dy/dx shall usually contain both x and y.

(c) *Parametric Equations:* Sometimes x and y are both expressed in terms of a third variable, say, t. This variable is called a parameter and the equations thus given are known as parametric equations. In the case of parametric equations $x = f_1(t)$ and $y = f_2(t)$, we find dy/dx in the following manner:

$$\frac{dy}{dx} = \frac{dy}{dt} \cdot \frac{dt}{dx} = \frac{dy/dt}{dx/dt}.$$

(d) *Trigonometrical Transformations:* Sometimes it is easy to differentiate after making some trigonometrical transformations.

Following formulae of trigonometry are of frequent use:

(i) $\cos x = 2 \cos^2 (x/2) - 1 = 1 - 2 \sin^2 (x/2)$;

(ii) $\sin x = 2 \tan \frac{1}{2} x \Big/ \left(1+\tan^2 \frac{1}{2} x\right)$;

(iii) $\cos x = \left(1-\tan^2 \frac{1}{2} x\right) \Big/ \left(1+\tan^2 \frac{1}{2} x\right)$;

(iv) $\tan x = 2\tan\frac{1}{2}x \Big/ \left(1-\tan^2\frac{1}{2}x\right)$;

(v) $\tan^{-1} x + \tan^{-1} y = \tan^{-1}\dfrac{x+y}{1-xy}$;

(vi) $\tan^{-1} x - \tan^{-1} y = \tan^{-1}\dfrac{x-y}{1+xy}$;

(vii) $2\tan^{-1} x = \tan^{-1} 2x/(1-x^2)$;

(viii) $3\tan^{-1} x = \tan^{-1}\dfrac{3x-x^3}{1+3x^2}$;

(ix) $\sin 3x = 3\sin x - 4\sin^3 x$;

(x) $\cos 3x = 4\cos^3 x - 3\cos x$.

SOLVED EXAMPLES

Example 1:

Find $\dfrac{dy}{dx}$ *if*

(i) $y = (x)^{\tan x} + (\sin x)^{\cos x}$

(ii) $y = (\sin x)^{\cos x} + (\cos x)^{\sin x}$

Solution:

(i) Let $u = (x)^{\tan x}$ and $v = (\sin x)^{\cos x}$

$\therefore$ $y = u + v \Rightarrow (dy/dx) = (du/dx) + (dv/dx)$ ...(1)

Now $u = (x)^{\tan x}$ $\log u = \tan x \log x$

$\therefore$ Differentiating w.r.t. x, we have

$(1/u)\,(du/dx) = (\sec^2 x)\,.\,\log x + (\tan x)/x$

or $\dfrac{du}{dx} = (x)^{\tan x}\left[\sec^2 x \log x + \dfrac{\tan x}{x}\right]$

Similarly, $\dfrac{dv}{dx} = (\sin x)^{\cos x}\,[(-\sin x)\,.\,\log \sin x + (\cos x)\,.\,\cot x]$.

$\therefore$ from (1), we get

$$\left(\frac{dy}{dx}\right) = \left(\frac{du}{dx}\right) + \left(\frac{dv}{dx}\right)$$

$$= (x)^{\tan x} [\sec^2 x \log x + (\tan x)/x]$$
$$+ (\sin x)^{\cos x} [\cos x \cot x - \sin x \,.\, \log \sin x].$$

(ii) Proceed exactly as in part (i).

Example 2:

Find (dy/dx), when

(i) $x = a (t - \sin t),\ y = a (1 - \cos t)$

(ii) $x = a (\cos t + \log \tan t/2),\ y = a \sin t$

(iii) $y = \tan^{-1} \dfrac{2t}{1-t^2},\ x = \sin^{-1} \dfrac{2t}{1+t^2}$

(iv) $x = a \sqrt{\left(\dfrac{t^2-1}{t^2+1}\right)},\ y = at\sqrt{\left(\dfrac{t^2-1}{t^2+1}\right)}$

Solution:

(i) We have $x = a (t - \sin t),\ y = a (1 - \cos t)$

$$\therefore \quad \frac{dx}{dt} = a (1 - \cos t) \text{ and } \frac{dy}{dt} = a \sin t$$

$$\text{Now } \frac{dy}{dx} = \frac{dy/dt}{dx/dt} = \frac{a \sin t}{a(1-\cos t)} = \frac{2 \sin t/2 \cos t/2}{2 \sin^2 t/2} = \cot t/2$$

(ii) Here $x = a (\cos t + \log \tan t/2),\ y = a \sin t$

$$\therefore \quad \frac{dx}{dt} = a \left[-\sin t + \frac{1}{\tan t/2} (\sec^2 t/2) \,.\, \frac{1}{2}\right]$$

$$= a\left(-\sin t + \frac{1}{2 \sin t/2 \cos t/2}\right)$$

$$= a\left[\frac{1}{\sin t} - \sin t\right] = a\left[\frac{1-\sin^2 t}{\sin t}\right] = \frac{a \cos^2 t}{\sin t}$$

Again $(dy/dt) = a \cos t$.

$$\text{Now } \frac{dy}{dx} = \frac{dy/dt}{dx/dt} = a \cos t \,.\, \frac{\sin t}{a \cos^2 t} = \tan t$$

(iii) Let $t = \tan \theta$. Then

$$y = \tan^{-1} \frac{2 \tan \theta}{1-\tan^2 \theta} = \tan^{-1} \tan 2\theta = 2\theta = 2 \tan^{-1} t$$

$\therefore \quad (dy/dt) = 2/(1 + t^2)$

Again $x = \sin^{-1} \dfrac{2t}{1+t^2} = \sin^{-1} \dfrac{2\tan\theta}{1+\tan^2\theta}$

$= \sin^{-1} (\sin 2\theta) = 2\theta = 2 \tan^{-1} t$

$\therefore \quad (dx/dt) = 2/(1 + t^2)$

Now $\dfrac{dy}{dx} = \dfrac{dy/dt}{dx/dt} = \dfrac{2/(1+t^2)}{2/(1+t^2)} = 1$

(iv) Here $x = a \sqrt{\left\{\dfrac{t^2-1}{t^2+1}\right\}}$ and $y = at \sqrt{\left\{\dfrac{t^2-1}{t^2+1}\right\}}$

We have, $\log y = \log a + \log t + \dfrac{1}{2} \log (t^2 - 1) = \dfrac{1}{2} \log (t^2 + 1)$

$$\therefore \quad \frac{1}{y}\frac{dy}{dt} = \frac{1}{t} + \frac{t}{t^2-1} - \frac{t}{t^2+1} = \frac{t^4 + 2t^2 - 1}{t(t^4-1)}$$

$$\text{or} \quad \sqrt{\left\{\frac{t^2-1}{t^2+1}\right\}} \cdot \frac{t^4+2t^2-1}{(t^4-1)}$$

Again $\log x = \log a + 1/2 \log (t^2 - 1) - 1/2 \log (t^2 + 1)$

$$\therefore \quad \frac{1}{x}\frac{dx}{dt} = \frac{t}{t^2-1} - \frac{1}{t^2+1} = \frac{2t}{t^4-1}$$

$$\text{or} \quad \frac{dx}{dt} = a\sqrt{\left(\frac{t^2-1}{t^2+1}\right)} \cdot \frac{2t}{t^4-1}$$

Now $\dfrac{dx}{dx} = \dfrac{dy/dt}{dx/dt} = \dfrac{t^4+2t^2-1}{2t}$.

Example 3(a):

Find $\dfrac{dy}{dx}$, *when* $x^y + (\sin x)^{\log x} = a$.

Solution:

Let $u = x^y$, $v = (\sin x)^{\log x}$ so that $u + v = a$

$$\therefore \quad \frac{du}{dx} + \frac{dv}{dx} = 0 \qquad ...(1)$$

We have log x = y log x and log v = log x log sin x

Differentiating w.r.t. 'x' we get

$$\frac{1}{u} \cdot \frac{dv}{dx} = \frac{y}{x} + \log x \frac{dy}{dx}$$

$$\frac{du}{dx} = x^y \left[\frac{y}{x} + \log x \frac{dy}{dx}\right] \qquad ...(2)$$

$$\text{Again } \frac{1}{v}\frac{dv}{dx} = \frac{1}{x} \log \sin x + \frac{\cos x}{\sin x} \log x$$

$$\frac{dv}{dx} = (\sin x)^{\log x} \left[\frac{1}{x} \log \sin x + \cot x \log x\right] \qquad ...(3)$$

Putting equations (2) and (3) in (1) we get

$$y\, x^{y-1} + x^y \log x \frac{dy}{dx} + (\sin x)^{\log x} \left[\frac{1}{x} \log \sin x + \cot x \log x\right] = 0$$

$$\text{Hence } \frac{dy}{dx} = \frac{(\sin x)^{\log x} \left[\frac{1}{x} \log \sin x + \cot x \log x\right] + yx^{y-1}}{x^y \log x}.$$

Example 3(b):

If $z = \sin\left(\frac{y}{x}\right)$, *then prove that* $\frac{\partial^2 z}{\partial x\, \partial y} = \frac{\partial^2 z}{\partial y\, \partial x}$.

Solution:

We have

$$\frac{\partial z}{\partial x} = \cos\left(\frac{y}{x}\right)\left(\frac{-y}{x^2}\right), \frac{\partial z}{\partial y} = \cos\left(\frac{y}{x}\right)\left(\frac{1}{x}\right)$$

$$\frac{\partial^2 z}{\partial x\, \partial y} = \frac{\partial}{\partial x}\left[\frac{\partial z}{\partial y}\right] = \frac{\partial}{\partial x}\left[\frac{1}{z}\cos\left(\frac{y}{x}\right)\right]$$

$$= -\frac{1}{x^2}\cos\left(\frac{y}{x}\right) - \frac{1}{x}\sin\left(\frac{y}{x}\right)\left(\frac{-y}{x^2}\right)$$

$$= -\frac{1}{x^2}\cos\left(\frac{y}{x}\right) + \frac{y}{x^2}\sin\left(\frac{y}{x}\right) \qquad ...(1)$$

$$\text{Now } \frac{\partial^2 z}{\partial y\, \partial x} = \frac{\partial}{\partial y}\left(\frac{\partial z}{\partial x}\right) = \frac{\partial}{\partial y}\left(-\frac{y}{x^2}\cos\frac{y}{x}\right)$$

$$= -\frac{1}{x^2}\left[1.\cos\left(\frac{y}{x}\right) - y\sin\left(\frac{y}{x}\right).\frac{1}{x}\right]$$

$$= -\frac{1}{x^2}\cos\left(\frac{y}{x}\right) + \frac{y}{x^2}\sin\left(\frac{y}{x}\right) \qquad ...(2)$$

from eqaution (1) and (2) we get

$$\frac{\partial^2 z}{\partial x\, \partial y} = \frac{\partial^2 z}{\partial y\, \partial x}.$$

Example 4:

Find (dy/dx) if $y = \tan^{-1}\{(ax - b)/(bx + a)\}$.

Solution:

We have $y = \tan^{-1}\left(\frac{ax - b}{bx + a}\right)$

$$\therefore \quad \frac{dy}{dx} = \frac{1}{1 + (ax - b)^2/(bx + a)^2} \cdot \frac{d}{dx}\left(\frac{ax - b}{bx + a}\right)$$

or
$$\frac{dy}{dx} = \frac{(bx + a)^2}{(bx - a)^2 + (ax - b)^2} \cdot \frac{a(bx + a) - b(ax - b)}{(bx + a)^2}$$

$$= \frac{a^2 + b^2}{(a^2 + b^2)x^2 + (a^2 + b^2)} = \frac{1}{1 + x^2}$$

Example 5:

If $z = f(y/x)$, *show that* $x\,(\partial z/\partial x) = +\, y(\partial z/\partial y) = 0$.

Solution:

$$\frac{\partial z}{\partial x} = \left[f'\left(\frac{y}{x}\right)\right]\left(-\frac{y}{x^2}\right), \text{ (diff. particularly w.r.t. x).}$$

$$\therefore \quad x\left(\frac{\partial z}{\partial x}\right) = -\left(\frac{y}{x}\right)f'\left(\frac{y}{x}\right). \qquad ...(1)$$

Again
$$\frac{\partial z}{\partial y} = \left[f'\left(\frac{y}{x}\right)\right]\left(\frac{1}{x}\right), \text{ (diff. partially w.r.t. y)}$$

$$\therefore \quad y\left(\frac{\partial z}{\partial y}\right) = \left(\frac{y}{x}\right)f'\left(\frac{y}{x}\right). \qquad ...(2)$$

Adding equations (1) and (2), we get $x\left(\frac{\partial z}{\partial x}\right) + y\left(\frac{\partial z}{\partial y}\right) = 0$.

Example 6(a):

If $u = \sin^{-1}\dfrac{x}{y} + \tan^{-1}\dfrac{y}{x}$, *show that* $\dfrac{\partial u}{\partial x} + y\dfrac{\partial u}{\partial y} = 0.$

Solution:

Here we have

$$\frac{\partial u}{\partial x} = \frac{1}{\sqrt{\left\{1-\left(\frac{x}{y}\right)^2\right\}}}.\frac{1}{y} + \frac{1}{1+\left(\frac{y}{x}\right)^2}.\left(-\frac{y}{x^2}\right), \text{ (treating as constant)}$$

$$= \frac{1}{\sqrt{(y^2-x^2)}} - \frac{1}{(x^2+y^2)}.$$

$$\therefore \quad x\frac{\partial u}{\partial x} = \frac{x}{\sqrt{(y^2-x^2)}} - \frac{xy}{x^2+y^2}. \qquad ...(1)$$

Again

$$\frac{\partial u}{\partial y} = \frac{1}{\sqrt{\left\{1-\left(\frac{x}{y}\right)^2\right\}}}\left(1-\frac{x}{y^2}\right) + \frac{1}{1+\left(\frac{y}{x}\right)^2}.\frac{1}{x}, \text{ (treating x as constant)}$$

$$= -\frac{1}{y\sqrt{(y^2-x^2)}} + \frac{x}{x^2+y^2}.$$

$$\therefore \quad y\frac{\partial u}{\partial y} = -\frac{r}{\sqrt{(y^2-x^2)}} + \frac{xy}{x^2+y^2}. \qquad ...(2)$$

Adding (1) and (2), we have $x\left(\dfrac{\partial u}{\partial x}\right) + y\left(\dfrac{\partial u}{\partial y}\right) = 0.$

Example 6(b):

Verify that $\dfrac{\partial^2 z}{\partial x\,\partial y} = \dfrac{\partial^2 z}{\partial y\,\partial x}$ *for the following functions:*

(i) $z = x \log xy$

(ii) $z = \sin^{-1}\left(\dfrac{x}{y}\right)$

(iii) $z = a\tan^{-1}\left(\dfrac{x}{y}\right)$

(iv) $z = x^y + y^x$

(v) $z = a^{ax} \sin by$

(vi) $z = \log (y \sin x + x \sin y)$.

Solution:

(i) We have $z = x \log y$.

Differentiating z partially w.r.t. a taking y as constant, we have

$$\frac{\partial z}{\partial x} = \log y.$$

Now $$\frac{\partial^2 z}{\partial y\, \partial x} = \frac{\partial}{\partial y}\left(\frac{\partial z}{\partial x}\right) = \frac{\partial}{\partial y}(\log y) = \frac{1}{y}.$$

Again differentiating z partially w.r.t. y taking x as constant, we have

$$\frac{\partial z}{\partial y} = \frac{x}{y}.$$

Now $$\frac{\partial^2 z}{\partial x\, \partial y} = \frac{\partial}{\partial x}\left(\frac{\partial z}{\partial y}\right) = \frac{\partial}{\partial x}\left(\frac{x}{y}\right) = \frac{1}{y},$$ treating y as constant.

Hence $$\frac{\partial^2 z}{\partial x\, \partial y} = \frac{\partial^2 z}{\partial y\, \partial x}.$$

(ii) We have $z = \sin^{-1}\left(\frac{x}{y}\right)$.

$$\therefore \quad \frac{\partial z}{\partial x} = \frac{1}{\sqrt{\left\{1-\left(\frac{x}{y}\right)^2\right\}}}.\frac{1}{y} = \frac{1}{\sqrt{(y^2 - x^2)}} = (y^2 - x^2)^{-1/2},$$

Now $$\frac{\partial^2 z}{\partial y\, \partial x} = \frac{\partial}{\partial y}\left(\frac{\partial z}{\partial x}\right) = \frac{\partial}{\partial y}\left\{(y^2 - x^2)^{-1/2}\right\}$$

$$= -\frac{1}{2}(y^2 - x^2)^{-3/2}.2y = \frac{-y}{(y^2 - x^2)^{3/2}}.$$

Again $$\frac{\partial z}{\partial y} = \frac{1}{\sqrt{\left\{1-\left(\frac{x}{y}\right)^2\right\}}}.\left(-\frac{x}{y^2}\right) = -\frac{x}{y}(y^2 - x^2)^{-1/2},$$

and $$\frac{\partial^2 z}{\partial x\, \partial y} = \frac{\partial}{\partial x}\left(\frac{\partial z}{\partial y}\right)$$

$$= -\frac{1}{y}\left(y^2 - x^2\right)^{-1/2} - \left(\frac{y}{x}\right).\left\{-\frac{1}{2}\left(y^2 - x^2\right)^{-3/2}\right\}.(-2x)$$

$$= -\frac{1}{y\left(y^2 - x^2\right)^{1/2}} - \frac{x^2}{y\left(y^2 - x^2\right)^{2/2}} = -\frac{\left(y^2 - x^2\right) + x^2}{y\left(y^2 - x^2\right)^{3/2}}$$

$$= \frac{-y}{\left(y^2 - x^2\right)^{3/2}}.$$

Hence $\dfrac{\partial^2 z}{\partial x\,\partial y} = \dfrac{\partial^2 z}{\partial y\,\partial x}$.

(iii) We have $z = a \tan^{-1}\left(\dfrac{x}{y}\right)$.

$$\therefore \quad \frac{\partial z}{\partial x} = a\frac{1}{1 + \left(\dfrac{x^2}{y^2}\right)}.\frac{1}{y} = \frac{ay}{x^2 + y^2},$$

and $\dfrac{\partial^2 z}{\partial y\,\partial x} = a.\dfrac{1.\left(x^2 + y^2\right) - y.2y}{\left(x^2 + y^2\right)^2} = \dfrac{a\left(x^2 + y^2\right)}{\left(x^2 + y^2\right)^2}$.

Again $\dfrac{\partial z}{\partial y} = a.\dfrac{1}{1 + \left(\dfrac{x^2}{y^2}\right)}.\left(-\dfrac{x}{y^2}\right) = \dfrac{-ax}{x^2 + y^2}$,

and $\dfrac{\partial^2 z}{\partial x\,\partial y} = -a.\dfrac{1.\left(x^2 + y^2\right) - x.2x}{\left(x^2 + y^2\right)^2} = -a.\dfrac{y^2 - x^2}{\left(x^2 + y^2\right)^2} = \dfrac{a\left(x^2 - y^2\right)}{\left(x^2 + y^2\right)^2}$

Thus, $\dfrac{\partial^2 z}{\partial x\,\partial y} = \dfrac{\partial^2 z}{\partial y\,\partial x}$.

(iv) We have $z = x^y + y^x$.

$$\therefore = \frac{\partial z}{\partial x} = yx^{y-1} + y^x.\log y$$

and $\dfrac{\partial^2 z}{\partial y\,\partial x} = \dfrac{\partial}{\partial y}\left[yx^{y-1} + y^x \log y\right]$

$$= 1.xy^{-1} + yx^{y-1}.\log x + xy^{x-1} \log y + y^x.\left(\frac{1}{y}\right)$$

$$= xy^{-1}(1 + y \log x) + y^{x-1}(1 + x \log y).$$

Again $\frac{\partial z}{\partial y}$ $x^y \log x + xy^{x-1}$ and $\frac{\partial^2 z}{\partial x\,\partial y} = \frac{\partial}{\partial x}$ $[x^y \log x + xy^{x-1}]$

$= yx^{y-1}.\log x + x^y.\left(\frac{1}{x}\right) + 1.y^{x-1} + xy^{x-1} \log y$

$= x^{y-1}(1 + y \log x) + y^{z-1}(1 + x \log y).$

Thus $\frac{\partial^2 z}{\partial x\,\partial y} = \frac{\partial^2 z}{\partial y\,\partial x}$.

(v) We have $z = e^{ax} \sin by$.

$\therefore \quad \frac{\partial z}{\partial x} = ae^{ax} \sin by,$

$\frac{\partial^2 z}{\partial y\,\partial x} = \frac{\partial}{\partial y}(ae^{ax} \sin by) = ae^{ax}.b \cos by = ab\, e^{ax} \cos by.$

Again $\frac{\partial z}{\partial y} = e^{ax}.b \cos by;$

$\frac{\partial^2 z}{\partial x\,\partial y} = \frac{\partial}{\partial x}(be^{ax} \cos by) = ab\, e^{ax} \cos by.$

Thus $\frac{\partial^2 z}{\partial y\,\partial x}\ \partial x = \frac{\partial^2 z}{\partial x\,\partial y}$.

(vi) We have $z = \log(y \sin x + x \sin y)$.

$\therefore \quad \frac{\partial z}{\partial x} = \frac{y \cos x + \sin y}{y \sin x + x \sin y},$

and $\frac{\partial^2 z}{\partial y\,\partial x}$

$$= \frac{(y \sin x + x \sin y)(\cos x + \cos y) - (y \cos x + \sin y)(\sin x + x \cos y)}{(y \sin x + x \sin y)^2}$$

$$= \frac{x \sin y \cos + y \sin x \cos y - \sin x \sin y - xy \cos x \cos y}{(y \sin x + x \sin y)^2}$$

Again $\frac{\partial z}{\partial y} = \frac{\sin x + x \cos y}{y \sin x + x \sin y}$

and $\frac{\partial^2 z}{\partial x\,\partial y}$

$$= \frac{(y \sin x + x \sin y)(\cos x + \cos y) - (\sin x + x \cos y)(y \cos x + \sin y)}{(y \sin x + x \sin y)^2}$$

$$= \frac{x \sin y \cos x + y \sin x \cos y - \sin x \sin y - xy \cos x \cos y}{(y \sin x + x \sin y)^2}.$$

$$\therefore \frac{\partial^2 z}{\partial x\, \partial y} = \frac{\partial^2 z}{\partial y\, \partial x}.$$

Example 6(c):

If $z = f(x + ay) + \phi\,(x - ay)$, prove that

$$\frac{\partial^2 z}{\partial y^2} = a^2\left(\frac{\partial^2 z}{\partial x^2}\right).$$

Solution:

We have $z = f(x + ay) + f(x - ay)$.

$$\therefore \quad \frac{\partial z}{\partial x} = f'(x + ay) + \phi'(x - ay), \text{ (diff. partially w.r.t. 'x')}$$

$$\text{and } \frac{\partial^2 z}{\partial x^2} = f''(x - y) + \phi''(x - y) \qquad \text{...(1)}$$

$$\text{Again } \frac{\partial z}{\partial y} = af'(x + ay) - a\phi'(x - ay).$$

$$\therefore \quad \frac{\partial^2 z}{\partial y^2} = a^2 f''(x + ay) + a^2\phi''(x - ay). \qquad \text{...(2)}$$

From (1) and (2), we get $\dfrac{\partial^2 z}{\partial y^2} = a^2\left(\dfrac{\partial^2 z}{\partial x^2}\right).$

Example 6(d):

If $z = xy\, f(y/x)$

prove that $x\dfrac{\partial z}{\partial x} + y\dfrac{\partial z}{\partial y} = 2z$

Solution:

$$\text{We have, } \frac{\partial z}{\partial x} = y\,[f(y/x) + xf'(y/x)\,(-y/x^2)]$$

$$\frac{\partial z}{\partial y} = x\,[f(y/x) + yf'(y/x)\,1/x]$$

$$\text{Hence, } x\frac{\partial z}{\partial x} + y\frac{\partial z}{\partial y} = xy\, f(y/x) = 22$$

Example 7:

If $z = \tan^{-1}\dfrac{x^2 + y^2}{x - y}$. Prove that

$$x\frac{\partial z}{\partial u}+y\frac{\partial z}{\partial y}=\sin 2z.$$

Solution:

We have

$$z=\tan^{-1}\frac{x^2+y^2}{x-y}$$

$$\Rightarrow \qquad \tan z=\frac{x^2+y^2}{x-y}$$

Let $\quad u=\tan z \qquad \text{...(1)}$

$$\therefore \qquad u=\frac{x^2+y^2}{x-y}=\frac{x^2\left\{1+\left(\frac{y}{x}\right)^2\right\}}{x\left\{1-\left(\frac{y}{x}\right)\right\}}=x^2f\left(\frac{y}{x}\right).$$

Thus, u is a homogeneous function of degree z.

Hence, the Euler's theorem;

$$x\frac{\partial u}{\partial x}+y\frac{\partial u}{\partial y}=2u \qquad \text{...(2)}$$

From (1) $\quad \frac{\partial u}{\partial u}=\sec^2 z\frac{\partial z}{\partial u},\ \frac{\partial u}{\partial y}=\sec^2 z\frac{\partial z}{\partial y} \qquad \text{...(3)}$

Putting (3) and (1) in (2) we get

$$\sec^2 z\left(x\frac{\partial z}{\partial x}+y\frac{\partial z}{\partial y}\right)=2\tan z$$

or

$$x\frac{\partial z}{\partial x}+y\frac{\partial z}{\partial y}=2\frac{\sin z}{\cos z}.\cos^2 z$$

$$2\sin z\cos z=\sin 2z.$$

Example 8(a):

If $z=x\,f\left(\frac{y}{x}\right)$ *prove that*

$$x\frac{\partial z}{\partial x}+y\frac{\partial z}{\partial y}=2z.$$

Solution:

We have $\quad \frac{\partial z}{\partial x}=y\left\{f\left(\frac{y}{x}\right)+xf'\left(\frac{y}{x}\right)\left(-\frac{y}{x^2}\right)\right\}$

$$\frac{\partial z}{\partial y} = x\left\{f\left(\frac{y}{x}\right) + yf'\left(\frac{y}{x}\right)\frac{1}{x}\right\}.$$

Hence $x\frac{\partial z}{\partial x} + y\frac{\partial z}{\partial y} = 2xy\, f\left(\frac{y}{x}\right) = 22.$

Example 8(b):

If $u = \sin^{-1}\left(\frac{x^2 + y^2}{x + y}\right)$, *show that*

$$x\left(\frac{\partial u}{\partial x}\right) + y\left(\frac{\partial u}{\partial y}\right) = \tan u.$$

Solution:

We have $\sin u = (x^2 + y^2)/(x + y)$.

$\therefore \quad \log \sin u = \log (x^2 + y^2) - \log (x + y)$. ...(1)

Differentiating (1) partially w.r.t. x, we get

$$\frac{1}{\sin u}\cos u.\frac{\partial u}{\partial x} = \frac{2x}{x^2 + y^2} - \frac{1}{x + y},$$

$$\therefore \quad (\cot u)\, x\frac{\partial u}{\partial x} = \frac{2x^2}{x^2 + y^2} - \frac{x}{x + y}. \quad ...(2)$$

Again differentiating (1) partially w.r.t. y, we get

$$\frac{\cos u}{\sin u}.\frac{\partial u}{\partial y} = \frac{2y}{x^2 + y^2} - \frac{1}{x + y}.$$

$$\therefore \quad (\cot u).y\frac{\partial u}{\partial y} = \frac{2y^2}{x^2 + y^2} - \frac{y}{x + y}. \quad ...(3)$$

Adding (2) and (3), we get

$$(\cot u)\left(x\frac{\partial u}{\partial x} + y\frac{\partial u}{\partial y}\right) = \frac{2x^2 + 2y^2}{x^2 + y^2} - \frac{x + y}{x + y} = 2 - 1 = 1.$$

$$\therefore \quad x\left(\frac{\partial u}{\partial x}\right) + y\left(\frac{\partial u}{\partial y}\right) = 1/\cot u = \tan u.$$

Example 8(c):

If $u = x^2 \tan^{-1}\frac{y}{x} - y^2 \tan^{-1}\frac{x}{y}$, *find* $\frac{\partial^2 u}{\partial x\, \partial y}$.

Solution:

We have

$$\frac{\partial u}{\partial y} = x^2 . \frac{1}{1+\left(\frac{y}{x}\right)^2} . \frac{1}{x} - 2y \tan^{-1}\frac{x}{y} - y^2 \frac{1}{1+\left(\frac{x}{y}\right)^2} . \left(-\frac{x}{y^2}\right)$$

$$= \frac{x^3}{x^2+y^2} - 2y \tan^{-1}\frac{x}{y} + \frac{xy^2}{x^2+y^2}$$

$$= \frac{x(x^2+y^2)}{x^2+y^2} - 2y \tan^{-1}\frac{x}{y} = x - 2y \tan^{-1}\frac{x}{y}.$$

Now $$\frac{\partial^2 u}{\partial x\, \partial y} = \frac{\partial}{\partial x}\left(\frac{\partial u}{\partial y}\right) = 1 - 2y \frac{1}{1+\left(\frac{x}{y}\right)^2} . \frac{1}{y} = 1 - \frac{2y^2}{x^2+y^2}$$

$$= \frac{x^2+y^2-2y^2}{x^2+y^2} = \frac{x^2-y^2}{x^2+y^2}.$$

Hence $$\frac{\partial^2 u}{\partial x\, \partial y} = \frac{x^2-y^2}{x^2+y^2}.$$

Example 9:

Find the value of

$$\frac{1}{a^2}\frac{\partial^2 z}{\partial x^2} + \frac{1}{b^2}\frac{\partial^2 z}{\partial y^2} \text{ when } a^2x^2 + b^2y^2 - c^2z^2 = 0.$$

Solution:

We have $$z^2 = \frac{a^2}{c^2}x^2 + \frac{b^2}{c^2}y^2. \qquad ...(1)$$

Differentiating (1) partially with respect to x taking y as constant, we have

$$2z\frac{\partial z}{\partial x} = 2\frac{a^2}{c^2}x \text{ or } \frac{\partial z}{\partial x} = \frac{a^2}{c^2}.\frac{x}{z}.$$

$$\therefore \quad \frac{\partial^2 z}{\partial x^2} = \frac{\partial}{\partial x}\left(\frac{a^2}{c^2}x.\frac{1}{z}\right) = \frac{a^2}{c^2}.\frac{1}{z} + \frac{c^2}{c^2}x.\left(-\frac{1}{z^2}\right)\frac{\partial z}{\partial x}$$

$$= \frac{a^2}{c^2 z} - \frac{a^2 x}{c^2 z^2}.\left(\frac{a^2 x}{c^2 z}\right) = \frac{a^2}{c^2 z} - \frac{a^4 x^2}{c^4 z^3}.$$

$$\therefore \quad \frac{1}{a^2}\frac{\partial^2 z}{\partial x^2} = \frac{1}{c^2 z} - \frac{a^2 x^2}{c^4 z^3} \qquad ...(2)$$

Again differentaiting (1) partially with respect to y taking x as constant, we have

$$2z\frac{\partial z}{\partial y} = 2\frac{b^2}{c^2}y \text{ or } \frac{\partial z}{\partial y} = \frac{b^2}{c^2}.\frac{y}{z}.$$

$$\therefore \quad \frac{\partial^2 z}{\partial y^2} = \frac{b^2}{c^2}.\frac{1}{z} + \frac{b^2}{c^2}y.\left(-\frac{1}{z^2}\right)\frac{\partial z}{\partial y}$$

$$= \frac{b^2}{c^2 z} - \frac{b^2 y}{c^2 z^2}.\left(\frac{b^2 y}{c^2 z}\right) = \frac{b^2}{c^2 z} - \frac{b^4 y^2}{c^4 z^3}.$$

$$\therefore \quad \frac{1}{b^2}\frac{\partial^2 z}{\partial y^2} = \frac{1}{c^2 z} - \frac{b^2 y^2}{c^4 z^3}. \quad ...(3)$$

Adding (2) and (3), we get

$$\frac{1}{a^2}\frac{\partial^2 z}{\partial x^2} + \frac{1}{b^2}\frac{\partial^2 z}{\partial y^2} = \frac{2}{c^2 z} - \frac{a^2 x^2 + b^2 y^2}{c^4 z^3}$$

$$= \frac{2}{c^2 z} - \frac{c^2 z^2}{c^4 z^3} - \frac{1}{c^2 z} - \frac{1}{c^2 z} = \frac{1}{c^2 z}.$$

Example 10(a):

If $z = xf(x + y) + y\phi\,(x + y)$*, prove that*

$$\left(\frac{\partial^2 z}{\partial x^2}\right) + \left(\frac{\partial^2 z}{\partial y^2}\right) = 2\left(\frac{\partial^2 z}{\partial x\,\partial y}\right).$$

Solution:

We have $z = xf(x + y) + y\phi(x + y)$.

$$\therefore \quad \frac{\partial z}{\partial x} = f(x + y) + xf'(x + y) + y\phi.(x + y),$$

$$\text{and } \frac{\partial^2 z}{\partial x^2} = f'(x + y) + f'(x + y) + xf''(x + y) + y\phi''(x + y)$$

$$= 2f'(x + y) + xf''(x + y) + y\phi''(x + y).$$

$$\text{Also } \frac{\partial y}{\partial z} = xf'(x + y) + \phi(x + y) + y\phi'(x + y)$$

$$\text{and } \frac{\partial^2 z}{\partial y^2} = xf''(x + y) + \phi'(x + y) + \phi'(x + y) + y\phi''(x + y)$$

$$= 2f'(x + y) + xf''(x + y) + yf''(x + y).$$

Again $\frac{\partial^2 z}{\partial x\, \partial y} = \frac{\partial}{\partial x}\left(\frac{\partial z}{\partial y}\right)$

$$= \frac{\partial}{\partial x}\{xf'(x+y) + \phi(x+y) + y\phi'(x+y)\}$$

$$= f'(x+y) + xf''(x+y) + \phi'(x+y) + y\phi''(x+y).$$

Now $\left(\frac{\partial^2 z}{\partial x^2}\right) + \left(\frac{\partial^2 x}{\partial y^2}\right)$

$$= 2[f'(x+y) + f''(x+y) + xf'(x+y) + y\phi''(x+y)]$$

$$= 2\left(\frac{\partial^2 z}{\partial x\, \partial y}\right).$$

Example 10(b):

If $u = tan^{-1}\frac{xy}{\sqrt{(1+x^2+y^2)}}$, *show that*

$$\frac{\partial^2 u}{\partial x\, \partial y} = \frac{1}{(1+x^2+y^2)^{3/2}}.$$

Solution:

We have

$$\frac{\partial u}{\partial x} = \frac{1}{1+\frac{x^2y^2}{1+x^2+y^2}}.$$

$$y.\frac{1.\sqrt{(1+x^2+y^2)} - x.\frac{1}{2}(1+x^2+y^2)^{-1/2}(-2x)}{1+x^2+y^2}$$

$$= \frac{1+x^2+y^2}{1+x^2+y^2+x^2y^2}.\frac{(1+x^2+y^2)-x^2}{(1+x^2+y^2)(1+x^2+y^2)^{1/2}}$$

$$= \frac{y(1+y^2)}{(1+x^2)(1+y^2)(1+x^2+y^2)^{1/2}} = \frac{1}{1+x^2}.\frac{y}{(1+x^2+y^2)^{1/2}}.$$

$\therefore$ $\frac{\partial^2 u}{\partial x\, \partial y} = \frac{\partial}{\partial y}\left(\frac{\partial u}{\partial x}\right)$

$$= \frac{1}{1+x^2} \cdot \frac{1.\left(1+x^2+y^2\right)^{1/2} - y.\frac{1}{2}\left(1+x^2+y^2\right)^{-1/2}.2y}{1+x^2+y^2}$$

$$= \frac{1}{1+x^2} \cdot \frac{1+x^2+y^2-y^2}{\left(1+x^2+y^2\right)\left(1+x^2+y^2\right)^{1/2}}$$

$$= \frac{1}{1+x^2} \cdot \frac{1+x^2}{\left(1+x^2+y^2\right)^{3/2}} = \frac{1}{\left(1+x^2+y^2\right)^{3/2}}.$$

Example 10(c):

If $\dfrac{x^2}{a^2+u} + \dfrac{y^2}{b^2+u} + \dfrac{z^2}{c^2+u} = 1$, *prove that*

$$\left(\frac{\partial u}{\partial x}\right)^2 + \left(\frac{\partial u}{\partial y}\right)^2 + \left(\frac{\partial u}{\partial z}\right)^2 = 2\left(x\frac{\partial u}{\partial x} + y\frac{\partial u}{\partial y} + z\frac{\partial u}{\partial z}\right).$$

Solution:

From the given equation we observe that u is a function of three independent variables x, y and z. Differentiating the given equation partially w.r.t. 'x', we get

$$\frac{2x}{a^2+u} - \left\{\frac{x^2}{\left(a^2+u\right)^2} + \frac{y^2}{\left(b^2+u\right)^2} + \frac{z^2}{\left(c^2+u\right)^2}\right\}\frac{\partial u}{\partial x} = 0.$$

$$\therefore \frac{\partial u}{\partial x} = \frac{2x/\left(a^2+u\right)}{\Sigma\left\{x^2/\left(a^2+u\right)^2\right\}}.$$

Similarly, by symmetry, we can write the values of $\dfrac{\partial u}{\partial y}$ and $\dfrac{\partial u}{\partial z}$.

$$\text{Now } \left(\frac{\partial u}{\partial x}\right)^2 = \frac{4x^2/\left(a^2+u\right)^2}{\left[\Sigma\left\{x^2/(a^2+u)^2\right\}\right]}.$$

$$\therefore \Sigma\left(\frac{\partial u}{\partial x}\right)^2 = \frac{4\Sigma\left\{x^2/\left(a^2+u\right)^2\right\}}{\left[\left\{x^2/\left(a^2+u\right)^2\right\}\right]}$$

$$= \frac{4}{\Sigma\left\{x^2/\left(a^2+u\right)^2\right\}} \qquad \text{...(1)}$$

Again $2x\dfrac{\partial u}{\partial x} = \dfrac{4x^2/(a^2+u)}{\Sigma\left\{x^2/(a^2+u)^2\right\}}$.

$$\therefore\ 2\Sigma x\left(\frac{\partial u}{\partial x}\right) = \frac{4\Sigma\left\{x^2/(a^2+u)\right\}}{\Sigma\left\{x^2/(a^2+u)^2\right\}} = \frac{4}{\Sigma\left\{x^2/(a^2+u)^2\right\}}. \qquad ...(2)$$

[$\because$ Σ $\{x^2/(a^2 + u)\} = 1$, from the given relation]

Now from (1) and (2), we have

$$\Sigma\left(\frac{\partial u}{\partial x}\right)^2 = 2\Sigma\left\{x\left(\frac{\partial u}{\partial x}\right)\right\}$$

Example 11:

If $z = (x^2 + y^2)/(x + y)$, *show that*

$$\left(\frac{\partial z}{\partial x} - \frac{\partial z}{\partial y}\right)^2 = 4\left(1 - \frac{\partial z}{\partial x} - \frac{\partial z}{\partial y}\right).$$

Solution:

We have $z = (x^2 + y^2)/(x + y)$.

$$\therefore\ \frac{\partial z}{\partial x} = \frac{(x+y).2x - (x^2+y^2).1}{(x+y)^2} = \frac{x^2 - y^2 + 2xy}{(x+y)^2}$$

and $\dfrac{\partial z}{\partial y} = \dfrac{(x+y)2y - (x^2+y^2).1}{(x+y)^2} = \dfrac{y^2 - x^2 + 2xy}{(x+y)^2}$.

$$\therefore\ \left(\frac{\partial z}{\partial x} - \frac{\partial z}{\partial y}\right)^2 = \left[\frac{(x^2 - y^2 + 2xy) - (y^2 - x^2 + 2xy)}{(x+y)^2}\right]^2$$

$$= \left[\frac{2(x^2 - y^2)^2}{(x^2 + y^2)}\right] = 4\left[\frac{x-y}{x+y}\right]^2.$$

Also $1 - \dfrac{\partial z}{\partial x} - \dfrac{\partial z}{\partial y} = 1 - \dfrac{x^2 - y^2 + 2xy}{(x+y^2)} - \dfrac{y^2 - x^2 + 2xy}{(x+y)^2}$

$$= \frac{(x^2 + y^2 + 2xy) - x^2 + y^2 - 2xy - y^2 + x^2 - 2xy}{(x+y)^2} = \frac{x^2 - 2xy + y^2}{(x+y)^2}$$

$$= \left(\frac{x-y}{x+y}\right)^2.$$

Hence $\left(\frac{\partial z}{\partial x}-\frac{\partial z}{\partial y}\right)^2 = 4\left(1-\frac{\partial z}{\partial x}-\frac{\partial z}{\partial y}\right).$

Example 12:

If $1/u = \sqrt{(x^2+y^2+z^2)}$, *show that*

$$x\frac{\partial u}{\partial x}+y\frac{\partial u}{\partial y}+z\frac{\partial u}{\partial z}=-u,$$

and $$\frac{\partial^2 u}{\partial x^2}+\frac{\partial^2 u}{\partial y^2}+\frac{\partial^2 u}{\partial z^2}=0.$$

Solution:

We have $u = (x^2 + y^2 + z^2)^{-1/2}$.

$\therefore \quad \frac{\partial u}{\partial x} = -\frac{1}{2}(x^2 + y^2 + z^2)^{-3/2}.(2x) = -x(x^2 + y^2 + z^2)^{-3/2},$

and $\frac{\partial^2 u}{\partial x^2} = x.\frac{3}{2}(x^2 + y^2 + z^2)^{-5/2}\ (2x) - 1(x^2 + y^2 + z^2)^{-3/2}$

$= (x^2 + y^2 + z^2)^{-5/2}\ [3x^2 - (x^2 + y^2 + z^2)]$

$= u^5(2x^2 - y^2 - z^2).$

Similarly, by symmetry $\frac{\partial u}{\partial y} = -y\ (x^2 + y^2 + z^2)^{-3/2},$

$\frac{\partial^2 u}{\partial y^2} = u^5\left(2y^2 - x^2 - z^2\right), \left(\frac{\partial u}{\partial z}\right) = -z(x^2 + y^2 + z^2)^{-3/2},$

Now $x\frac{\partial u}{\partial x}+y\frac{\partial u}{\partial y}+z\frac{\partial u}{\partial z}$

$= -x^2(x^2 + y^2 + z^2)^{-3/2} - y^2(x^2 + y^2 + z^2)^{-3/2} - z^2(x^2 + y^2 + z^2)^{-3/2}$

$= -(x^2 + y^2 + z^2)^{-3/2}\ (x^2 + y^2 + z^2) = -(x^2 + y^2 + z^2)^{-1/2} = -u.$

Also $\frac{\partial^2 u}{\partial x^2}+\frac{\partial^2 u}{\partial y^2}+\frac{\partial^2 u}{\partial z^2}$

$= u^5\ (2x^2 - y^2 - z^2 + 2y^2 - x^2 - z^2 - x^2 - y^2) = u^5.0 = 0.$

Example 13:

If $u = \log r$, *where* $r^2 = (x - a)^2 + (y - b^2) + (z - c)^2$, *show that*

$$\frac{\partial^2 u}{\partial x^2}+\frac{\partial^2 u}{\partial y^2}+\frac{\partial^2 u}{\partial z^2}=\frac{1}{r^2}.$$

Solution:

We have $r^2 = (x - a)^2 + (y - b)^2 + (z - c)^2$. ...(1)

Differentiating (1) partially wrt x we have

$$2r\frac{\partial r}{\partial x}=2(x-a) \text{ or } \frac{\partial r}{\partial x}=\frac{1}{r}\left(\frac{x-a}{r}\right) \quad ...(2)$$

Now $u = \log r$. $\therefore \frac{\partial u}{\partial x}=\frac{1}{r}$

$$\therefore \frac{\partial r}{\partial x}=\frac{1}{r}\left(\frac{x-a}{r}\right). \quad ...[\text{from } (2)]$$

Thus, $\frac{\partial u}{\partial x}=\frac{(x-a)}{r^2}$

$$\therefore \frac{\partial^2 u}{\partial x^2}=\frac{\partial}{\partial x}\left(\frac{\partial u}{\partial x}\right)=\frac{\partial}{\partial x}\left(\frac{x-a}{r^2}\right)=\frac{r^2(1)-(x-a).2r\left(\frac{\partial r}{\partial x}\right)}{r^4}$$

$$=\frac{r^2-2(x-a)^2}{r^4}, \quad \left[\because \text{from } (2), \frac{\partial r}{\partial x}=(x-a)/r\right]$$

Similarly, by symmetry

$$\frac{\partial^2 u}{\partial y^2}=\frac{r^2-2(y-b)^2}{r^4} \text{ and } \frac{\partial^2 u}{\partial z^2}=\frac{r^2-2(z-c)^2}{r^4}.$$

Hence $\frac{\partial^2 u}{\partial x^2}+\frac{\partial^2 u}{\partial y^2}+\frac{\partial^2 u}{\partial z^2}=\frac{3r^2-2\left\{(x-a)^2+(y-b)^2+(z-c)^2\right\}}{r^4}$

$$=\frac{3r^2-2r^2}{r^4}, \{\text{using } (1)\} = \frac{r^2}{r^4}=\frac{1}{r^2}.$$

Example 14:

If $u = e^{xyz}$, show that

$$\frac{\partial^3 u}{\partial x\, \partial y\, \partial z} = (1 + 3xyz + x^2y^2z^2)e^{xyz}.$$

Solution:s

Here $u = e^{xyz}$. $\therefore \frac{\partial u}{\partial z}= xy\, e^{xyz}$.

$$\text{Now } \frac{\partial^2 u}{\partial y\,\partial z} = \frac{\partial}{\partial y}\left(\frac{\partial u}{\partial z}\right) = \frac{\partial}{\partial y}\left(xy\,e^{xyz}\right) = x\frac{\partial}{\partial y}\left(y\,e^{xyz}\right)$$

$$= x[y.xz\,e^{xyz} + e^{xyz}] = e^{xyz}\,(x^2yz + z).$$

$$\text{Again } \frac{\partial^3 u}{\partial x\,\partial y\,\partial z} = \frac{\partial}{\partial x}\left(\frac{\partial^2 u}{\partial y\,\partial z}\right) = \frac{\partial}{\partial x}\left[e^{xyz}\left(x^2yz + x\right)\right]$$

$$= e^{xyz}\,(2xyz + 1) + yz\,e^{xyz}(x^2yz + x)$$

$$= e^{xyz}\,[2xyz + 1 + x^2y^2z^2 + xyz]$$

$$= e^{xyz}\,[1 + 3xyz + x^2y^2z^2].$$

Example 15:

If $u \log(x^2 + y^2 + z^2)$, *show that*

$$x\frac{\partial^2 u}{\partial y\,\partial z} = y\frac{\partial^2 u}{\partial z\,\partial x} = z\frac{\partial^2 u}{\partial x\,\partial y}.$$

Solution:

We have $u = \log(x^2 + y^2 + z^2)$.

$$\therefore \frac{\partial u}{\partial z} = \frac{2z}{x^2 + y^2 + z^2}, \text{ [treating x and y as constants]}$$

$$\text{and } \frac{\partial^2 u}{\partial y\,\partial z} = \frac{\partial}{\partial y}\left(\frac{\partial u}{\partial z}\right) = 2z.\left[-\left(x^2 + y^2 + z^2\right)^{-2}.2y\right]$$

$$= -\,4yz/(x^2 + y^2 + z^2)^2.$$

$$\text{Now } x\frac{\partial^2 u}{\partial y\,\partial z} = -\,4xyz/(x^2 + y^2 + z^2)^2.$$

$$\text{By symmetry, } y\frac{\partial^2 u}{\partial z\,\partial x} = z\frac{\partial^2 u}{\partial x\,\partial y} = -\frac{4xyz}{\left(x^2 + y^2 + z^2\right)^2}.$$

$$\text{Hence } x\frac{\partial^2 u}{\partial y\,\partial z} = y\frac{\partial^2 u}{\partial z\,\partial x} = z\frac{\partial^2 u}{\partial x\,\partial y}.$$

Example 16:

If $x^x\,y^y\,z^z = c$ *show that* $x = y - z$,

$$\frac{\partial^2 z}{\partial x\,\partial y} = -\,\{x \log(ex)\}^{-1}.$$

Solution:

We have $x^x.y^y.z^z = c$. From this equation we observe that we can regard z as a function of two independent variables x and y. Taking logarithms of both sides of the given equation, we get

$$x \log x + y \log y + z \log z = \log c. \qquad ...(1)$$

Now differentiating (1) partially w.r.t. x taking y as constant, we have

$$x.\frac{1}{x} + 1.\log x + \left[z.\frac{1}{z} + 1.\log z\right]\frac{\partial z}{\partial x} = 0.$$

[Note that z is not a constant but is a function of x and y]

$$\therefore \qquad \frac{\partial z}{\partial x} = -\frac{(1+\log x)}{(1+\log z)}. \qquad ...(2)$$

Similarly differentiating (1) partially w.r.t. y, we have

$$\frac{\partial z}{\partial y} = -\frac{(1+\log y)}{(1+\log z)}. \qquad ...(3)$$

Now $\dfrac{\partial^2 z}{\partial x\, \partial y} = \dfrac{\partial}{\partial x}\left(\dfrac{\partial z}{\partial y}\right) = \dfrac{\partial}{\partial x}\left[-\left(\dfrac{1+\log y}{1+\log z}\right)\right]$, [from (3)]

$$= -(1+\log y).\frac{\partial}{\partial x}[(1+\log z)^{-1}]$$

$$= -(1+\log y).\left[-(1+\log z)^{-2}.\frac{1}{z}.\frac{\partial z}{\partial x}\right]$$

$$= \frac{(1+\log y)}{z(1+\log z)^2}.\left[-\left(\frac{1+\log x}{1+\log z}\right)\right], \qquad \text{[from (2)]}.$$

Hence, when x = y = z, we have

$$\frac{\partial^2 z}{\partial x\, \partial y} = -\frac{(1+\log x)^2}{x(1+\log x)^3}, \quad \left[\text{putting } y = z = x \text{ in the value of } \left(\frac{\partial^2 z}{\partial x\, \partial y}\right)\right]$$

$$= -\frac{1}{x(1+\log x)} = -\frac{1}{x(\log e + \log x)}, \qquad [\because \log e = 1]$$

$$= -\frac{1}{x\log(ex)} = -\{x \log(ex)\}^{-1}.$$

Example 17(a):

If $u = \log(x^3 + y^3 + z^3 - 3xyz)$, show that

$$\frac{\partial u}{\partial x} + \frac{\partial u}{\partial y} + \frac{\partial u}{\partial z} = \frac{3}{x+y+z}$$

and $$\left(\frac{\partial}{\partial x} + \frac{\partial}{\partial y} + \frac{\partial}{\partial z}\right)^2 u = \frac{-9}{(x+y+z)^2}.$$

Solution:

We have $u = \log(x^3 + y^3 + z^3 - 3xyz)$.

$$\therefore \quad \frac{\partial u}{\partial x} = \frac{3x^2 - 3yz}{x^3 + y^3 + z^3 - 3xyz}, \quad \frac{\partial u}{\partial y} = \frac{3y^2 - 3zx}{x^3 + y^3 + z^3 - 3xyz}$$

and $$\frac{\partial u}{\partial z} = \frac{3z^2 - 3xy}{x^3 + y^3 + z^3 - 3xyz}$$

$$\therefore \quad \frac{\partial u}{\partial x} + \frac{\partial u}{\partial y} + \frac{\partial u}{\partial z} = \frac{3(x^2 + y^2 + z^2 - yz - zx - xy)}{x^3 + y^3 + z^3 - 3xyz}$$

$$= \frac{3(x^2 + y^2 + z^2 - yz - zx - xy)}{(x+y+z)(x^2 + y^2 + z^2 - yz - zx - xy)} = \frac{3}{x+y+z}. \quad \ldots(1)$$

Now $$\left(\frac{\partial}{\partial x} + \frac{\partial}{\partial y} + \frac{\partial}{\partial z}\right)^2 u = \left(\frac{\partial}{\partial x} + \frac{\partial}{\partial y} + \frac{\partial}{\partial z}\right)\left(\frac{\partial u}{\partial x} + \frac{\partial u}{\partial y} + \frac{\partial u}{\partial z}\right)$$

$$= \left(\frac{\partial}{\partial x} + \frac{\partial}{\partial y} + \frac{\partial}{\partial z}\right)\left(\frac{3}{x+y+z}\right), \quad \text{from (1)}$$

$$= 3\left[\frac{\partial}{\partial x}\left(\frac{1}{x+y+z}\right) + \frac{\partial}{\partial y}\left(\frac{1}{x+y+z}\right) + \frac{\partial}{\partial z}\left(\frac{1}{x+y+z}\right)\right]$$

$$= 3\left[\frac{-1}{(x+y+z)^2} + \frac{-1}{(x+y+z)^2} + \frac{-1}{(x+y+z)^2}\right] = \frac{-9}{(x+y+z)^2}.$$

Example 17(b):

If $z = \tan(y + ax) + (y - ax)^{3/2}$, find the value of

$$\left(\frac{\partial^2 z}{\partial x^2}\right) - a^2\left(\frac{\partial^2 z}{\partial y^2}\right).$$

Solution:

Here $z = \tan(y + ax) + (y - ax)^{3/2}$.

$$\therefore \quad \left(\frac{\partial z}{\partial x}\right) = \{\sec^2(y+ax)\}.a + \frac{3}{2}(y-ax)^{1/2}.(-a),$$

and $$\left(\frac{\partial^2 z}{\partial x^2}\right) = 2a^2 \tan(y+ax)\sec^2(y+ax) + \frac{3}{4}a^2(y-ax)^{-1/2}.$$

Again $$\left(\frac{\partial z}{\partial y}\right) = \sec^2(y+ax) + \frac{3}{2}(y-ax)^{1/2}$$

and $\left(\frac{\partial^2 z}{\partial y^2}\right) = 2\sec^2(y+ax)\tan(y+ax) + \frac{3}{4}(y-ax)^{-1/2}.$

Thus, $\left(\frac{\partial^2 z}{\partial x^2}\right) - a^2\left(\frac{\partial^2 z}{\partial y^2}\right) = 0.$

Example 18:

If $u = f(r)$ where $r^2 = x^2 + y^2$, show that

$$\frac{\partial^2 u}{\partial x^2} + \frac{\partial^2 u}{\partial y^2} = f''(r) + \frac{1}{r}f'(r).$$

Solution:

Differentiating $r^2 = x^2 + y^2$ partially w.r.t. x and y, we get

$$2r\frac{\partial r}{\partial x} = 2x \text{ or } \frac{\partial r}{\partial x} = \frac{x}{r};\ 2r\frac{\partial r}{\partial y} = 2y \text{ or } \frac{\partial r}{\partial y} = \frac{y}{r}. \quad ...(1)$$

Now $u = f(r)$. Therefore $\frac{\partial u}{\partial x} = \{f'(r)\} = \frac{x}{r}f'(r)$ [from (1)]

and $\frac{\partial^2 u}{\partial x^2} = \frac{\partial}{\partial x}\left(\frac{\partial u}{\partial x}\right) = \frac{\partial}{\partial x}\left[x.\frac{1}{r}.f'(r)\right]$

$$= 1.\frac{1}{r}.f'(r) + \{xf'(r)\}\left(-\frac{1}{r^2}\frac{\partial r}{\partial x}\right) + \frac{x}{r}\{f''(r)\}\frac{\partial r}{\partial x}$$

$$= \frac{1}{r}f'(r) - \frac{x}{r^2}.\frac{x}{r}f'(r) + \frac{x^2}{r^2}f''(r), \quad \left[\because \text{from (1), } \frac{\partial x}{\partial r} = \frac{x}{r}\right]$$

$$= \left(\frac{1}{r}\right)f'(r) - \left(\frac{x^2}{r^3}\right)f'(r) + \left(\frac{x^2}{r^2}\right)f''(r). \quad ...(2)$$

Similarly, by symmetry, we have

$$\frac{\partial^2 u}{\partial y^2} = \frac{1}{r}f'(r) - \frac{y^2}{r^3}f'(r) + \frac{y^2}{r^2}f''(r). \quad ...(3)$$

Adding (2) and (3), we get

$$\frac{\partial^2 u}{\partial x^2} + \frac{\partial^2 u}{\partial y^2} = \frac{2}{r}f'(r) - \frac{x^2+y^2}{r^3}f'(r) + \frac{x^2+y^2}{r^2}f''(r)$$

$$= \left(\frac{2}{r}\right)f'(r) - \left(\frac{r^2}{r^3}\right)f'(r) + \left(\frac{r^2}{r^2}\right)f''(r), \quad [\because r^2 = x^2 + y^2]$$

$$=\left(\frac{2}{r}\right)f'(r)-\left(\frac{1}{r}\right)f'(r)+f''(r)=f''(r)+\left(\frac{1}{r}\right)f'(r).$$

Example 19(a):

If $u = x\phi\,(y/x) = \psi(y/x)$, prove that

$$x^2\frac{\partial^2 u}{\partial x^2}+2xy\frac{\partial^2 u}{\partial x\,\partial y}+y^2\frac{\partial^2 u}{\partial y^2}=0.$$

Solution:

We have $u = x\phi\,(y/x) + \psi\,(y/x)$. ...(1)

Differentiating (1) partially w.r.t. x and y, we get

$$\left(\frac{\partial u}{\partial x}\right)=x\left\{\phi'\left(\frac{y}{x}\right)\right\}.\left(-\frac{y}{x^2}\right)+\phi\left(\frac{y}{x}\right)+\left\{\psi'\left(\frac{y}{x}\right)\right\}.\left(-\frac{y}{x^2}\right),$$

and $$\left(\frac{\partial u}{\partial x}\right)=x\left\{\phi'\left(\frac{y}{x}\right)\right\}.\left(\frac{1}{x}\right)+\phi\left(\frac{y}{x}\right)+\left\{\psi'\left(\frac{y}{x}\right)\right\}.\left(\frac{1}{x}\right).$$

$$\therefore\; x\left(\frac{\partial u}{\partial x}\right)+y\left(\frac{\partial u}{\partial y}\right)=x\phi\left(\frac{y}{x}\right)$$ ...(2)

Now differentiating (2) partially w.r.t. x and y respectively, we get

$$x\frac{\partial^2 u}{\partial x^2}+\frac{\partial u}{\partial y}+y\frac{\partial^2 u}{\partial x\,\partial x}=x\left\{\phi'\left(\frac{y}{x}\right)\right\}\left(-\frac{y}{x^2}\right)+\phi\left(\frac{y}{x}\right),$$

and $$x\frac{\partial^2 u}{\partial x\,\partial y}+y\frac{\partial^2 u}{\partial y^2}+\frac{\partial u}{\partial y}=x\left\{\phi'\left(\frac{y}{x}\right)\right\}.\frac{1}{x}.$$

Multiplying these equations by x and y respectively and adding, we get

$$x^2\frac{\partial^2 u}{\partial x^2}+2xy\frac{\partial^2 u}{\partial x\,\partial y}+y^2\frac{\partial^2 u}{\partial y^2}+x\frac{\partial u}{\partial x}+y\frac{\partial u}{\partial y}=x\phi\left(\frac{y}{x}\right)$$

or $$x^2\frac{\partial^2 u}{\partial x^2}+2xy\frac{\partial^2 u}{\partial x\,\partial y}+y^2\frac{\partial^2 u}{\partial y^2}=0,$$ [from (2)]

Example 19(b):

If $x = r\cos\theta$, $y = r\sin\theta$, prove that

(a) $$\frac{\partial^2 r}{\partial x^2}+\frac{\partial^2 r}{\partial y^2}=\frac{1}{r}\left[\left(\frac{\partial r}{\partial x}\right)^2+\left(\frac{\partial r}{\partial y}\right)^2\right],$$

(b) $$\frac{\partial^2 r}{\partial x^2}.\frac{\partial^2 r}{\partial y^2}=\left(\frac{\partial^2 r}{\partial x\,\partial y}\right)^2,$$

(c) $\left(\frac{\partial r}{\partial x}\right)^2 + \left(\frac{\partial r}{\partial y}\right)^2 = 1.$

Solution:

(a) We have x = r cos θ, y = r sin θ.

Therefore $r^2 = x^2 + y^2$ ...(1)

Now $2r\left(\frac{\partial r}{\partial x}\right) = 2x;$ [diff. (1) partially w.r.t 'x']

$$\therefore \quad \frac{\partial r}{\partial x} = \frac{x}{r}.$$

$$\frac{\partial^2 r}{\partial x^2} = \frac{r.1 - x.\frac{\partial r}{\partial x}}{r^2} = \frac{r - x.\frac{x}{r}}{r^2}, \qquad \text{...[using (2)]}$$

$$= \frac{r^2 - x^2}{r^3} = \frac{(x^2 + y^2) - x^2}{r^3} = \frac{y^2}{r^3}. \qquad \text{...(3)}$$

Again differentiating (1) partially w.r.t. 'y', we get

$$2r\frac{\partial r}{\partial y} = 2y; \qquad \therefore \frac{\partial r}{\partial y} = \frac{y}{r} \qquad \text{...(4)}$$

Differentiating (4) partially w.r.t. y, we get

$$\frac{\partial^2 r}{\partial y} = \frac{r.1 - y.\frac{\partial r}{\partial y}}{r^2} = \frac{r - y.\frac{y}{r}}{r^2} \qquad \left[\because \text{from (4)}, \frac{\partial r}{\partial r} = \frac{y}{r}\right]$$

$$= \frac{r^2 - y^2}{r^3} = \frac{(x^2 + y^2) - y^2}{r^3} = \frac{x^2}{r^3}. \qquad \text{...(5)}$$

Adding (3) and (5), we get

$$\frac{\partial^2 r}{\partial x^2} + \frac{\partial^2 r}{\partial y^2} = \frac{y^2}{r^3} + \frac{x^2}{r^3} = \frac{x^2 + y^2}{r^3} = \frac{1}{r}.$$

Also $\frac{1}{r}\left[\left(\frac{\partial r}{\partial x}\right)^2 + \left(\frac{\partial r}{\partial y}\right)^2\right] = \frac{1}{r}\left[\frac{x^3}{r^2} + \frac{y^2}{r^2}\right] = \frac{x^2 + y^2}{r^3} = \frac{r^2}{r^3} = \frac{1}{r}.$

$$\therefore \frac{\partial^2 r}{\partial x^2} + \frac{\partial^2 r}{\partial y^2} = \frac{1}{r}\left[\left(\frac{\partial r}{\partial x}\right)^2 + \left(\frac{\partial r}{\partial y}\right)^2\right].$$

(b) Differentiating (4) partially w.r.t. 'x', we get

$$\frac{\partial^2 r}{\partial x\,\partial y} = -\frac{y}{r^2}.\frac{\partial r}{\partial x} = -\frac{xy}{r^3}, \qquad \left[\because \text{ from (2)}, \frac{\partial r}{\partial x} = \frac{x}{r}\right]$$

Now $\dfrac{\partial^2 r}{\partial x^2}.\dfrac{\partial^2 r}{\partial y^2} = \dfrac{y^2}{r^3}.\dfrac{x^2}{r^3}$, [from (3) and (5)]

$$= \frac{x^2y^2}{r^6} = \left(\frac{-xy}{r^3}\right)^2 = \left(\frac{\partial^2 r}{\partial x\,\partial y}\right)^2.$$

(c) From (2) and (4), on squaring and adding, we get

$$\left(\frac{\partial r}{\partial x}\right)^2 + \left(\frac{\partial r}{\partial y}\right)^2 = \frac{x^2}{r^2} + \frac{y^2}{r^2} = \frac{x^2+y^2}{r^2} = \frac{r^2}{r^2} = 1.$$

Example 19(c):

If $u = (1 - 2xy + y^2)^{-1/2}$, *prove that*

$$\frac{\partial}{\partial x}\left\{(1-x^2)\frac{\partial u}{\partial x}\right\} + \frac{\partial}{\partial y}\left\{y^2\frac{\partial u}{\partial y}\right\} = 0.$$

Solution:

Here $u = (1 - 2xy + y^2)^{-1/2}$.

$$\therefore \qquad \frac{\partial u}{\partial x} = -\frac{1}{2}(1-2xy+y^2)^{-3/2}(-2y) = yu^3$$

and $\qquad \dfrac{\partial u}{\partial y} = -1/2\ (1 - 2xy + y^2)^{-3/2}\ (-2y + 2y) = (x - y)\ u^3.$

Now $\qquad \dfrac{\partial}{\partial x}\left\{(1-x^2)\dfrac{\partial u}{\partial x}\right\} = \dfrac{\partial}{\partial x}\left\{(1-x^2).yu^3\right\}$

$$= y(-2x)u^3 + y(1 - x^2).3u^2\frac{\partial u}{\partial x}$$

$$= -2xy\ u^3 + 3y\ (1 - x^2)u^2.yu^3$$

$$= -2xy\ u^3 + 3y^2u^5\ (1 - x)^2. \qquad \text{...(1)}$$

Also $\dfrac{\partial}{\partial y}\left\{y^2\dfrac{\partial u}{\partial y}\right\} = \dfrac{\partial}{\partial y}\left\{y^2(x-y)u^3\right\} = \dfrac{\partial}{\partial y}\left\{(y^2x-y^3)u^3\right\}$

$$= (2xy - 3y^2)\ u^3 + (y^2x - y^3).3u^2\frac{\partial u}{\partial y}$$

$$= (2xy - 3y^2)\ u^3 + y^2(x - y).3u^2.(x - y)u^3$$

$$= (2xy - 3y^2)u^3 + y^2(x - y)^2.3u^5$$

$$= 2xy\ u^3 + 3y^2u^5\ [(x - y)^2 - u^{-2}]$$

Example 20:

If $u \tan^{-1} \dfrac{x^3+y^3}{x-y}$, *find the value of*

$$x^2\frac{\partial^2 u}{\partial x^2}+2xy\frac{\partial^2 u}{\partial x\,\partial y}+y^2\frac{\partial^2 u}{\partial y^2}.$$

Solution:

Thus we get

$$x\left(\frac{\partial u}{\partial x}\right)+y\left(\frac{\partial u}{\partial y}\right)=\sin 2u \qquad ...(1)$$

Now differentiating (1) partially w.r.t. x and y respectively, we get

$$x\frac{\partial^2 u}{\partial x^2}+\frac{\partial u}{\partial x}+y\frac{\partial^2 u}{\partial x\,\partial y}=2\cos 2u.\frac{\partial u}{\partial x}, \qquad ...(2)$$

and $x\dfrac{\partial^2 u}{\partial x\,\partial y}+\dfrac{\partial u}{\partial y}+y\dfrac{\partial^2 u}{\partial y^2}=2\cos 2u\dfrac{\partial u}{\partial y}.$ ...(3)

Multiplying equations (2) by x, (3) by y and adding, we get

$$x^2\frac{\partial^2 u}{\partial x^2}+2xy\frac{\partial^2 u}{\partial x\,\partial y}+y^2\frac{\partial^2 u}{\partial y^2}+x\frac{\partial u}{\partial x}+y\frac{\partial u}{\partial y} = 2\cos 2u\left(x\frac{\partial y}{\partial x}+y\frac{\partial u}{\partial y}\right)$$

or $x^2\dfrac{\partial^2 u}{\partial x^2}+2xy\dfrac{\partial^2 u}{\partial x\,\partial y}+y^2\dfrac{\partial^2 u}{\partial y^2} = \sin 2u = 2\cos 2u \sin 2u$, [by (1)]

or $x^2\dfrac{\partial^2 u}{\partial x^2}+2xy\dfrac{\partial^2 u}{\partial x\,\partial y}+y^2\dfrac{\partial^2 u}{\partial y^2} = \sin 4u - \sin 2u.$

Example 21(a):

If $z = \tan^{-1}\left(\dfrac{x+y}{\sqrt{x}+\sqrt{y}}\right)$ *then*

$$x\frac{\partial z}{\partial x}+y\frac{\partial z}{\partial y}=\frac{1}{4}\sin 2z.$$

Solution:

$$z = \tan^{-1}\left(\frac{x+y}{\sqrt{x}+\sqrt{y}}\right) \Rightarrow \tan z = \frac{x+y}{\sqrt{x}+\sqrt{y}}.$$

Let u = tan z ...(1)

u is a homogeneous functions in x and y of degree $\frac{1}{2}$.

By Euler's Theorem $x\frac{\partial y}{\partial x}+y\frac{\partial y}{\partial y}=\frac{1}{2}$.

or $$x\left(\sec^2 z\frac{\partial z}{\partial x}\right)+y\left(\sec^2 z\frac{\partial z}{\partial y}\right)=\frac{1}{2}\tan x \text{ by (1)}$$

or $$x\frac{\partial z}{\partial x}+y\frac{\partial z}{\partial y}=\frac{1}{2}\frac{\tan z}{\sec^2 z}=\frac{1}{2}\sin z\cos z$$

$$=\frac{1}{4}\sin 2z.$$

Example 21(b):

If f(x, y, z) is a homogeneous function of the nth degree in x, y, z, prove that

$$x^2\frac{\partial^2 f}{\partial x^2}+y^2\frac{\partial^2 f}{\partial y^2}+z^2\frac{\partial^2 f}{\partial z^2}+2yz\frac{\partial^2 f}{\partial y\,\partial z}+2zx\frac{\partial^2 f}{\partial z\,\partial x}+2xy\frac{\partial^2 f}{\partial x\,\partial y}$$
$$= n(n-1)\,f(x, y, z).$$

Solution:

Here f(x, y, z) is a homogeneous function of the nth degree n, x, y, z.

Therefore $\frac{\partial f}{\partial x}$, $\frac{\partial f}{\partial y}$ and $\frac{\partial f}{\partial z}$ are homogeneous functions f the (n – 1)th degree in x, y, z. So using Euler's theorem for $\frac{\partial f}{\partial x}$,

$$x\frac{\partial}{\partial x}\left(\frac{\partial f}{\partial x}\right)+y\frac{\partial}{\partial y}\left(\frac{\partial f}{\partial x}\right)+z\frac{\partial}{\partial z}\left(\frac{\partial f}{\partial x}\right)=(n-1)\frac{\partial f}{\partial x}$$

or $$x\frac{\partial^2 f}{\partial x^2}+y\frac{\partial^2 f}{\partial y\,\partial x}+z\frac{\partial^2 f}{\partial z\,\partial x}=(n-1)\frac{\partial f}{\partial x}. \quad ...(1)$$

Similarly, $$x\frac{\partial^2 f}{\partial x\,\partial y}+y\frac{\partial^2 f}{\partial y^2}+z\frac{\partial^2 f}{\partial y\,\partial z}=(n-1)\frac{\partial f}{\partial y} \quad ...(2)$$

and $$x\frac{\partial^2 f}{\partial x\,\partial y}+y\frac{\partial^2 f}{\partial z^2}=(n-1)\frac{\partial f}{\partial z} \quad ...(3)$$

Multiplying (1) by x, (2) by y and (3) by z and adding, we get

$$x^2\frac{\partial^2 f}{\partial x^2}+y^2\frac{\partial^2 f}{\partial y^2}+z^2\frac{\partial^2 f}{\partial z^2}+2yz\frac{\partial^2 f}{\partial y\,\partial z}+2zx\frac{\partial^2 f}{\partial z\,\partial x}+2xy\frac{\partial^2 f}{\partial x\,\partial y}$$

$$= (n-1)\left(x\frac{\partial f}{\partial y} + \frac{\partial f}{\partial y}z\frac{\partial f}{\partial z}\right)$$

$= (n - 1)\, nf(x, y, z) = n\,(n - 1)\, f(x, y, z)$.

Example 21(c):

Verify Euler's theorem in the following cases:

(i) $u = x^{-4}\, 3x^3y + 5x^2y^2 + 4xy^3 - 2y^4$.

(ii) $u = \dfrac{x(x^3 - y^3)}{x^3 + y^3}$,

(iii) $u = \dfrac{x^{1/4} + y^{1/4}}{x^{1/5} + y^{1/5}}$,

(iv) $u = axy + byz = czx$,

(v) $u = x^n \log\left(\dfrac{y}{x}\right)$,

(vi) $u = 1/\sqrt{(x^2 + y^2)}$.

Solution:

(i) We have $= x^4 - 3x^3y + 5x^2y^2 + 4xy^3 - 2y^4$. Obviously u is a homogeneous function of x and y of degree 4. So by Euler's Theorem,

we must have $x\left(\dfrac{\partial u}{\partial x}\right) + y\left(\dfrac{\partial u}{\partial y}\right) = 4u$. Let us verify it.

We have $\left(\dfrac{\partial u}{\partial x}\right) = 4x^3 - 9x^2y + 10xy^2 + 4y^3$,

and $\left(\dfrac{\partial u}{\partial y}\right) = -3x^3 + 10x^2y + 12xy^2 - 8y^3$.

$$\therefore \quad x\frac{\partial u}{\partial x} + y\frac{\partial u}{\partial y} = x(4x^3 - 9x^2y + 10xy^2 + 4y^3) + y(-3x^3 + 10x^2y + 12xy^2 - 8y^3)$$

$= 4(x^4 - 3x^3y + 5x^2y^2 + 4xy^3 - 2y^4)$

$= 4u$. This verifies Euler's Theorem.

(ii) We have $u = \dfrac{x(x^3 - y^3)}{x^3 + y^3}$ which is obviously a homogeneous function of x and y of degree 4 – 3 *i.e.*, 1. Note that each term in the numerator is of degree 4 while each term in the denominator is of degree 3. In order to verify Euler's Theorem we are to know that

$$x\frac{\partial u}{\partial x}+y\frac{\partial u}{\partial y}=1.u=u.$$

Now $\log u = \log x + \log (x^3 - y^3) - \log(x^3 + y^3)$. ...(1)

Differentiating (1) partially w.r.t. x and y respectively, we get

$$\frac{1}{u}\frac{\partial u}{\partial x}=\frac{1}{x}+\frac{3x^2}{x^3-y^3}-\frac{3x^2}{x^3+y^3} \qquad ...(2)$$

and $$\frac{1}{u}\frac{\partial u}{\partial y}=0-\frac{3y^2}{x^3-y^3}-\frac{3y^2}{x^3+y^3}. \qquad ...(3)$$

Multiplying (2) by x and (3) by y and adding, we get

$$\frac{1}{u}\left(x\frac{\partial u}{\partial x}+y\frac{\partial u}{\partial y}\right)=1+\frac{3(x^3-y^3)}{x^3-y^3}-\frac{3(x^3+y^3)}{x^3+y^3}$$

$= 1 + 3 - 3 = 1.$

$\therefore\ x\frac{\partial u}{\partial x}+y\frac{\partial u}{\partial y}=u$. This verifies Euler's Theorem.

(iii) Here u is a homogeneous function of x and y of degree $\frac{1}{4}-\frac{1}{5}$ *i.e.*, $\frac{1}{20}$. So by Euler's Theorem we must have

$$x\left(\frac{\partial u}{\partial x}\right)+y\left(\frac{\partial u}{\partial y}\right)=\frac{1}{20}u.$$

Let us verify it. We have

$\log u = \log(x^{1/4} + y^{1/4}) - \log (x^{1/5} + y^{1/5}).$

$$\therefore\ \frac{1}{u}\frac{\partial u}{\partial x}=\frac{\frac{1}{4}x^{-3/4}}{x^{1/4}+y^{1/4}}-\frac{\frac{1}{5}x^{-4/5}}{x^{1/5}+y^{1/5}}$$

and $$\frac{1}{u}\frac{\partial u}{\partial y}=\frac{\frac{1}{4}y^{-3/4}}{x^{1/4}+y^{1/4}}-\frac{\frac{1}{5}y^{-4/5}}{x^{1/5}+y^{1/5}}.$$

$$\therefore\ \frac{1}{u}\left(x\frac{\partial u}{\partial x}+y\frac{\partial u}{\partial y}\right)=\frac{1}{4}\frac{x^{1/4}+y^{1/4}}{x^{1/4}+y^{1/4}}-\frac{1}{5}\frac{x^{1/5}+y^{1/5}}{x^{1/5}+y^{1/5}}=\frac{1}{4}-\frac{1}{5}=\frac{1}{20}.$$

$\therefore\ x\frac{\partial u}{\partial x}+y\frac{\partial u}{\partial y}=\frac{1}{20}u.$ This verifies Euler's Theorem.

(iv) We have = axy + byz + czx, which is a homogeneous function of x, y and z of degree 2. So in order to verify Euler's Theorem, we must show that $x\frac{\partial u}{\partial x}+y\frac{\partial u}{\partial y}+z\frac{\partial u}{\partial z}=2u$.

Now $\frac{\partial u}{\partial x}=ay+cz,\ \frac{\partial u}{\partial y}=ax+bz,$ and $\frac{\partial u}{\partial z}=by+cx$.

$\therefore\ x\frac{\partial u}{\partial x}+y\frac{\partial u}{\partial y}+z\frac{\partial u}{\partial z}=x\,(ay+cz)+y\,(ax+bz)+z\,(by+cx)$

$= 2\,(axy + byz + czx) = 2u$. This verifies Euler's Theorem.

(v) Here u is a homogeneous function of x and y of degree n. So by Euler's Theorem we must have

$$x\left(\frac{\partial u}{\partial x}\right)+y\left(\frac{\partial u}{\partial y}\right)=nu.$$

Now do the verification yourself.

(vi) Here $u=\frac{1}{\sqrt{(x^2+y^2)}}=\frac{1}{x\sqrt{[1+(y/x^2)]}}=x^{-1}.\frac{1}{\sqrt{\left[1+\sqrt{(y/x)^2}\right]}}$ is a homogeneous function of x and y of degree – 1. Now proceed yourself.

Example 22:

If $u=\tan^{-1}\left(\frac{x^3+y^3}{x+y}\right)$, *show that*

$$x\left(\frac{\partial u}{\partial x}\right)+y\left(\frac{\partial u}{\partial y}\right)=\sin 2u.$$

Solution:

We have tan u = $(x^3 + y^3)/(x + y)$ = v, say. Then v is a homogeneous function of x and y of degree 3 – 1 *i.e.*, 2. Therefore by Euler's theorem.

We have $x\left(\frac{\partial v}{\partial x}\right)+y\left(\frac{\partial v}{\partial y}\right)=2v.$ ...(1)

Now v = tan u.

$\therefore\quad \frac{\partial v}{\partial x}=\sec^2 u\frac{\partial u}{\partial x}$ and $\frac{\partial v}{\partial y}=\sec^2 u\frac{\partial u}{\partial y}$.

Putting these values in (1), we get

$$x \sec^2 u \frac{\partial u}{\partial x} + y \sec^2 u \frac{\partial u}{\partial y} = 2v$$

or $$x\frac{\partial u}{\partial x} + y\frac{\partial u}{\partial y} = \frac{2v}{\sec^2 u} = \frac{2\tan u}{\sec^2 u} = 2 \sin u \cos u = \sin 2u.$$

Example 23:

Prove that $\dfrac{d^2y}{dx^2} + \dfrac{2a^2x^2}{y^5} = 0$,

where $y^3 - 3ax^2 + x^3 = 0$.

Solution:

Let $f(x, y) \equiv y^3 - 3ax^2 + x^3 = 0$.

Then $p = \dfrac{\partial f}{\partial x} = -6ax + 3x^2$, $q = \dfrac{\partial f}{\partial y} = 3y^2$,

$$r = \frac{\partial^2 f}{\partial x^2} = -6a + 6x,\ s = \frac{\partial^2 f}{\partial x\,\partial y} = 0,\ t = \frac{\partial^2 f}{\partial y^2} = 6y.$$

Now $\dfrac{d^2y}{dx^2} = -\dfrac{q^2r - 2pqs + p^2t}{q^3}$,

$$= -\frac{(-6a+6x)\left(3y^2\right)^2 + 6y\left(-6x+3x^2\right)^2}{\left(3y^2\right)^3}$$

$$= -\frac{2(-a+x)y^3 + 2\left(-2ax+x^2\right)^2}{y^5}$$

$$= -\frac{2x\left(y^3+x^3-3ax^2\right) - 2ay^3 + 8a^2x^2 - 2ax^3}{y^5}$$ **(Note)**

$$= -\frac{-2a\left(y^3+x^3\right) + 8a^2x^2}{y^5},$$ $[\because y^3 + x^3 - 3ax^2 = 0]$

$$= -\frac{-2a\left(3ax^2\right) + 8a^2x^2}{y^5} = -\frac{2a^2x^2}{y^5}.$$

Hence $\dfrac{d^2y}{dx^2} + \dfrac{2a^2x^2}{y^5} = 0$.

Example 24:

If $(\tan x)^y + y^{\cot x} = a$, *find* $\dfrac{dy}{dx}$.

Solution:

Let $f(x, y) = (\tan x)^y + y^{\cot x} - a$. Then

$$\left(\frac{\partial f}{\partial x}\right) = y\,(\tan x)^{y-1}.\sec^2 x + y^{\cot x}.\log y.(-\operatorname{cosec}^2 x),$$

and $$\left(\frac{\partial f}{\partial y}\right) = (\tan x)^y \log \tan x + (\cot x).y^{\cot x - 1}$$

Now we are given that $f(x, y) = 0$.

$$\therefore\ \frac{dy}{dx} = -\frac{\partial f/\partial x}{\partial f/\partial y} = -\frac{y(\tan x)^{y-1}\sec^2 x - y^{\cot x}.\log y.\operatorname{cosec}^2 x}{(\tan x)^y \log \tan x + \cot x.y^{\cot x-1}}.$$

Example 25(a):

If $x^3y^3 + 3x \sin y = e^y$, find $\dfrac{dy}{dx}$.

Solution:

Let $f(x, y) = x^3y^3 + 3x \sin y - e^x$. Then we have $f(x, y) = 0$.

$$\therefore\ \frac{dy}{dx} = -\frac{\partial f/\partial x}{\partial f/\partial y} = -\frac{\left(3x^2y^3 + 3\sin y\right)}{3x^3y^2 + 3x\cos y - e^y}.$$

Example 25(b):

If $x^y + y^x = a^b$, find $\dfrac{dy}{dx}$.

Solution:

Let $f(x, y) = x^y + y^x - a^b$. Then we have $f(x, y) = 0$.

$$\therefore\ \frac{dy}{dx} = -\frac{\partial f/\partial x}{\partial f/\partial y} = \frac{yx^{y-1} + y^x \log y}{x^y \log x + xy^{x-1}}.$$

Example 26:

If $u = \log\{(x^2 + y^2)/xy\}$, find du.

Solution:

We have $u = \log(x^2 + y^2) - \log x - \log y$.

$$\therefore\ \frac{\partial u}{\partial x} = \frac{2x}{x^2+y^2} - \frac{1}{x} = \frac{2x^2 - x^2 - y^2}{x\left(x^2+y^2\right)} = \frac{x^2-y^2}{x\left(x^2+y^2\right)},$$

and $$\frac{\partial u}{\partial y} = \frac{2y}{x^2+y^2} - \frac{1}{y} = \frac{2y^2 - x^2 - y^2}{y\left(x^2+y^2\right)} = \frac{\left(y^2-x^2\right)}{y\left(x^2+y^2\right)}.$$

Now $du = \frac{\partial u}{\partial x}dx + \frac{\partial u}{\partial y}dy = \frac{(x^2 - y^2)}{x(x^2 + y^2)}dx + \frac{(y^2 - x^2)}{y(x^2 + y^2)}dy$

$$= \frac{x^2 - y^2}{xy(x^2 + y^2)}\,(y\,dx - x\,dy).$$

Example 27:

If $u = \sin(x^2 + y^2)$, where $a^2x^2 + b^2y^2 = c^2$, find $\frac{du}{dx}$.

Solution:

We have $\frac{du}{dx} = \frac{\partial u}{\partial x} + \frac{\partial u}{\partial y}.\frac{dy}{dx}$. ...(1)

Now $u = \sin(x^2 + y^2)$.

$\therefore\ \frac{\partial u}{\partial y} = 2x\cos(x^2 + y^2)$ and $\frac{\partial u}{\partial y} = 2y\cos(x^2 + y^2)$.

Since $a^2x^2 + b^2y^2 = c^2$,

Therefore $2a^2x + 2b^2y\left(\frac{dy}{dx}\right) = 0$ or $\frac{dy}{dx} = -\frac{(a^2x)}{(b^2y)}$.

$\therefore$ from (1), $\frac{du}{dx} = 2x\cos(x^2 + y^2) - [2y\cos(x^2 + y^2)]\left[\frac{(a^2x)}{(b^2y)}\right]$

$$= [2x\cos(x^2 + y^2)]\left[1 - \left(\frac{a^2}{b^2}\right)\right].$$

Example 28(a):

If $z = xy\,f\left(\frac{y}{x}\right)$, show that $x\left(\frac{\partial z}{\partial x}\right) + y\left(\frac{\partial z}{\partial y}\right) = 2z$.

Show also that if z is a constant,

$$\frac{f'\left(\frac{y}{x}\right)}{f\left(\frac{y}{x}\right)} = \frac{x\left\{y + x\left(\frac{dy}{dx}\right)\right\}}{y\left\{y - x\left(\frac{dv}{dx}\right)\right\}}.$$

Solution:

We have, $z = x^2.\left(\frac{y}{x}\right)f\left(\frac{y}{x}\right)$, so that z is a homogeneous function of x and y of degree 2.

Hence by Euler's Theorem, we have

$$x\left(\frac{\partial z}{\partial x}\right)+y\left(\frac{\partial z}{\partial y}\right)=2z.$$

If z be a constant, then differentiating

$z = xy\, f\left(\frac{y}{x}\right)$ logarithmically, w.r.t. x, we get

$$0=\frac{1}{x}+\frac{1}{y}\frac{dy}{dx}+\frac{f'\left(\frac{y}{x}\right)}{f\left(\frac{y}{x}\right)}\cdot\frac{x\left(\frac{dy}{dx}\right)-y}{x^2}$$

$$=\frac{y+x\left(\frac{dy}{dx}\right)}{xy}+\frac{f'\left(\frac{y}{x}\right)}{f\left(\frac{y}{x}\right)}\cdot\frac{x\left(\frac{dy}{dx}\right)-y}{x^2}$$

Hence $\dfrac{f'\left(\frac{y}{x}\right)}{f\left(\frac{y}{x}\right)}=\dfrac{x\left\{y+x\left(\frac{dy}{dx}\right)\right\}}{y\left\{y-x\left(\frac{dy}{dx}\right)\right\}}$.

Example 28(b):

If $u = \log\dfrac{x^3+y^3}{x+y}$, *show that* $x\dfrac{\partial u}{\partial x}+y\dfrac{\partial u}{\partial y}=2$.

Solution:

We have $e^u = \dfrac{x^3+y^3}{x+y} = v$, say.

Obviously $v = (x^3 + y^3)/(x + y)$ is a homogeneous function of x and y of degree 3 – 1 *i.e.*, 2. Therefore by Euler's theorem, we have

$$x\frac{\partial v}{\partial x}+y\frac{\partial v}{\partial y}=2.v=2v \qquad ...(1)$$

Now $v = e^u$.

$\therefore \dfrac{\partial v}{\partial x}=e^u\dfrac{\partial u}{\partial x}$ and $\dfrac{\partial v}{\partial y}=e^u\dfrac{\partial u}{\partial y}$.

Putting these values in (1), we get

$$xe^{u}\frac{\partial u}{\partial x}+y\,e^{u}\frac{\partial u}{\partial y}=2e^{u}$$

or $e^{u}\left(x\frac{\partial u}{\partial x}+y\frac{\partial u}{\partial y}\right)=2e^{u}$ or $x\frac{\partial u}{\partial x}+y\frac{\partial u}{\partial y}=2$.

Example 28(c):

If $u = x\phi\left(\frac{y}{x}\right)+\psi\left(\frac{y}{x}\right)$, *prove that*

$$x^2\frac{\partial^2 u}{\partial x^2}+2xy\frac{\partial^2 u}{\partial x\,\partial y}+y^2\frac{\partial^2 u}{\partial y^2}=0.$$

Solution:

Let $u = z_1 + z_2$, where $z_1 = x\phi\left(\frac{y}{x}\right)$ and $z_2 = \psi\left(\frac{y}{x}\right)$. Obviously z_1 is a homogeneous function of x and y of degree 1 and z_2 is a homogeneous function of x and y of degree zero. Now

$$x\frac{\partial u}{\partial x}+y\frac{\partial u}{\partial y}=x\frac{\partial}{\partial x}(z_1+z_2)+y\frac{\partial}{\partial y}(z_1+z_2)$$

$$=\left(x\frac{\partial z_1}{\partial x}+y\frac{\partial z_1}{\partial y}\right)+\left(x\frac{\partial z_2}{\partial x}+y\frac{\partial z_2}{\partial y}\right)=1.z_1+0.z_2.\text{ (by Euler's theorem).}$$

Thus, $x\left(\frac{\partial u}{\partial x}\right)+y\left(\frac{\partial u}{\partial y}\right)=z_1$ ...(1)

Differentiating equation (1) partially w.r.t. x and y respectively, we get

$$x\frac{\partial^2 u}{\partial x^2}+\frac{\partial u}{\partial x}+y\frac{\partial^2 u}{\partial x\,\partial y}=\frac{\partial z_1}{\partial x}, \quad \text{...(2)}$$

and $x\frac{\partial^2 u}{\partial x\,\partial y}+\frac{\partial u}{\partial y}+y\frac{\partial^2 u}{\partial y^2}=\frac{\partial z_1}{\partial y}$. ...(3)

Multiplying (2) by x and (3) by y and adding, we get

$$x^2\frac{\partial^2 u}{\partial x^2}+2xy\frac{\partial^2 u}{\partial x\,\partial y}+y^2\frac{\partial^2 u}{\partial y^2}+x\frac{\partial u}{\partial x}+y\frac{\partial u}{\partial y}=x\frac{\partial z_1}{\partial x}+y\frac{\partial z_1}{\partial y}$$

or $$x^2\frac{\partial^2 u}{\partial x^2}+2xy\frac{\partial^2 u}{\partial x\,\partial y}+y^2\frac{\partial^2 u}{\partial y^2}+z_1=1.z_1,$$

$$\left[\because x\left(\frac{\partial u}{\partial x}\right)+y\left(\frac{\partial u}{\partial y}\right)=z_1 \text{by (1), and } x\left(\frac{\partial z_1}{\partial x}\right)+y\left(\frac{\partial z_1}{\partial y}\right)=1.z_1 \text{ by Euler's theorem}\right]$$

or $$x^2\frac{\partial^2 u}{\partial x^2}+2xy+y^2\frac{\partial^2 u}{\partial y^2}=0.$$

Example 29:

If $u = x^2y^2/(x + y)$, *show that*

$$x\left(\frac{\partial u}{\partial x}\right)+y\left(\frac{\partial u}{\partial y}\right)=3u.$$

Solution:

We have $u=\dfrac{x^2y^2}{x+y^2}=\dfrac{x^3\left(\frac{y}{x}\right)^2}{\left[1+\left(\frac{y}{x}\right)\right]}=x^3f\left(\frac{y}{x}\right)$, say.

Thus u is a homogeneous function of x and y of degree 3.

Therefore by Euler's theorem, we have $x\dfrac{\partial u}{\partial x}+y\dfrac{\partial u}{\partial y}=3u$.

Example 30:

If $u = \sin^{-1}\left(\dfrac{x+y}{\sqrt{x}+\sqrt{y}}\right)$, *show that*

$$x\left(\frac{\partial u}{\partial x}\right)+y\left(\frac{\partial u}{\partial y}\right)=\frac{1}{2}\tan u.$$

Solution:

We have $\sin u = (x+y)/(\sqrt{x}+\sqrt{y}) = v$, say.

Then v is a homogeneous function of x and y of degree $\left(1-\frac{1}{2}\right)$ *i.e.,* $\frac{1}{2}$.

Applying Euler's Theorem for v, we have

$$x\left(\frac{\partial v}{\partial x}\right)+y\left(\frac{\partial v}{\partial y}\right)=\frac{1}{2}v$$

$$\text{or } x\frac{\partial}{\partial x}(\sin u)+y\frac{\partial}{\partial y}(\sin u)=\frac{1}{2}\sin u, \qquad [\because v = \sin u]$$

$$\text{or } x\cos u\left(\frac{\partial u}{\partial x}\right)+y\cos u\left(\frac{\partial u}{\partial y}\right)=\frac{1}{2}\sin u$$

$$\text{or } x\left(\frac{\partial u}{\partial x}\right)+y\left(\frac{\partial u}{\partial y}\right)=\frac{1}{2}\tan u.$$

Example 31:

If u be a homogeneous function of x and y of degree n, show that

$$x\left(\frac{\partial^2 u}{\partial x^2}\right)+y\left(\frac{\partial^2 u}{\partial x\,\partial y}\right)=(n-1)\left(\frac{\partial u}{\partial x}\right),$$

$$\text{and } x\left(\frac{\partial^2 u}{\partial x\,\partial y}\right)+y\left(\frac{\partial^2 u}{\partial y^2}\right)=(n-1)\left(\frac{\partial u}{\partial y}\right).$$

Hence deduce that

$$x^2\frac{\partial^2 u}{\partial x^2}+2xy\frac{\partial^2 u}{\partial x\,\partial y}+y^2\frac{\partial^2 u}{\partial y^2}=n(n-1)\,u.$$

Solution:

By Euler's Theorem, we have

$$x\left(\frac{\partial u}{\partial x}\right)+y\left(\frac{\partial u}{\partial y}\right)=nu \qquad ...(1)$$

Differentiating (1) partially w.r.t. x, we get

$$x\frac{\partial^2 u}{\partial x^2}+\frac{\partial u}{\partial x}+y\frac{\partial^2 u}{\partial x\,\partial y}=n\frac{\partial u}{\partial x}$$

$$\text{or } x\frac{\partial^2 u}{\partial x^2}+y\frac{\partial^2 u}{\partial x\,\partial y}=n\frac{\partial u}{\partial x}-\frac{\partial u}{\partial x}=(n-1)\frac{\partial u}{\partial x} \qquad ...(2)$$

Similarly, differentiating (1) partially w.r.t. 'y', we get

$$x\frac{\partial^2 u}{\partial x\,\partial y}+y\frac{\partial^2 u}{\partial y^2}=(n-1)\frac{\partial u}{\partial y}. \qquad ...(3)$$

Now multiplying (2) by x and (3) by y and adding, we get

$$x^2\frac{\partial^2 u}{\partial x^2}+2xy\frac{\partial^2 u}{\partial x\,\partial y}+y^2\frac{\partial^2 u}{\partial y^2}$$

$$= (n-1)\left[x\frac{\partial u}{\partial x}+y\frac{\partial u}{\partial y}\right] = (n-1)\,nu = n(n-1)\upsilon.$$

Example 32:

If $u = \sin^{-1}\left\{\left(\sqrt{x}-\sqrt{y}\right)/\left(\sqrt{x}+\sqrt{y}\right)\right\}$, *show that*

$$\frac{\partial u}{\partial x} = -\frac{y}{x}\frac{\partial u}{\partial y}.$$

Solution:

We have $\sin u = \left(\sqrt{x}-\sqrt{y}\right)/\left(\sqrt{x}+\sqrt{y}\right) = v$, say. Then v is a homogeneous function of x and y of degree $\frac{1}{2}-\frac{1}{2}$ *i.e.*, 0.

Therefore by Euler's Theorem, we have $x\frac{\partial v}{\partial x}+y\frac{\partial v}{\partial y}=0.\ v = 0.$ ...(1)

Now v = sin u.

$$\therefore \frac{\partial v}{\partial x} = \cos u\frac{\partial u}{\partial x}, \frac{\partial v}{\partial y} = \cos u\frac{\partial u}{\partial y}.$$

Putting these values in (1), we get

$$x\cos u\frac{\partial u}{\partial x}+y\cos u\frac{\partial u}{\partial y}=0 \text{ or } \cos u\left(x\frac{\partial u}{\partial x}+y\frac{\partial u}{\partial y}\right)=0$$

or $x\frac{\partial u}{\partial x}+y\frac{\partial u}{\partial y} = 0$, since $\cos u \neq 0$. Hence $\frac{\partial u}{\partial x} = -\frac{y}{x}\frac{\partial u}{\partial y}$.

EXERCISES

1. Verify Euler's Theorem for

$$z = \tan^{-1}\left(\frac{y}{x}\right).$$

2. It $u = \sin^{-1}\frac{x^2+y^2}{x+y}$ show that

$$x\frac{\partial u}{\partial x}+y\frac{\partial u}{\partial y} = \tan u.$$

3. If $z = \tan^{-1}\left(\frac{x^2+y^2}{x+y}\right)$ find $\frac{\partial z}{\partial x}+\frac{\partial z}{\partial y}$ and $\frac{\partial^2 z}{\partial x\,\partial y}$.

4. If $x^x y^y z^z = c$, show that x = y = z

$$\frac{\partial^2 z}{\partial x\,\partial y} = -(x\log ex)^{-1}.$$

5. If $u \tan^{-1}\frac{xy}{\sqrt{1+x^2+y^2}}$ show that

$$\frac{\partial^2 u}{\partial x\, \partial y} = \frac{1}{\left(1 + x^2 + y^2\right)^{3/2}}.$$

6. If $u = e^{xyz}$ show that

$$\frac{\partial^2 u}{\partial x\, \partial y\, \partial z} = (1 + 3xyz + x^2y^2z^2)e^{xyz}.$$

7. If $u = \sin^{-1}\left(\frac{\sqrt{x} - \sqrt{y}}{\sqrt{x} + \sqrt{y}}\right)$ show that $x\frac{\partial u}{\partial x} + y\frac{\partial u}{\partial y} = 0$.

8. If $z = x\left(\frac{y}{x}\right)$ prove that $x\frac{\partial z}{\partial x} + y\frac{\partial z}{\partial y} = z$.

9. If $u \tan^{-1}\left(\frac{x + y}{\sqrt{x} + \sqrt{y}}\right)$ prove that

$$x^2\frac{\partial^2 u}{\partial x^2} + 2xy\frac{\partial^2 y}{\partial x\, \partial y} + y^2\frac{\partial^2 y}{\partial y^2} = \frac{\sin 2u}{8}\left\{2\cos^2 u - 3\right\}.$$

10. If $u = At^{-1/2}\, e^{-x^2/4a^2t}$ prove that

$$\frac{\partial u}{\partial t} = a^2\frac{\partial^2 u}{\partial x^2}.$$

11. Show that the function $z = \sin\left(\frac{y}{z}\right)$ $\frac{\partial^2 z}{\partial y\, \partial x} = \frac{\partial^2 z}{\partial x\, \partial y}$.

12. If $u = (x^2 + y^2 + z^2)^{-3/2}$ show that $x\frac{\partial u}{\partial x} + y\frac{\partial u}{\partial y} + z\frac{\partial u}{\partial z} = -4$.

6

Application of Differential Equations

ORDINARY AND PARTIAL DIFFERENTIAL EQUATIONS

In most of the cases, from the available data, we get such relationships involving derivatives. In analyzing any system, one of the basic problems is to determine the functional relationship among variables. Assume that the independent variable x and the dependent variable y are related in such a way that, the differential coefficient of y with respect to x doubles for every unit increase of x. This can be written in equation form as:

$$\frac{dy}{dx} = 2 \qquad \ldots(1)$$

Can we find a relation between x and y which satisfies (1)? By inspection we see the equation $y = x^2$ satisfies (1). Hence $y = x^2$ is 'a solution' of the equation. We used the term 'a solution', since we do not know whether it has any other solution. Further, we get

$$y = x^2 + 1,\ y = x^2 - \frac{1}{2},\ \text{or}\ y = x^2 + 2$$

which are all solutions of (1). In general, $y = x^2 + c$, for any constant c is a solution of the equation. For each value of c, $y = x^2 + c$ represents a curve in the xy-plane. In other words, $y = x^2 + c$ represents a family of curves, for different values of c. Equation (1) is called a *differential equation.*

Thus, a differential equation is an equation that involves derivatives of one or more unknown functions. Differential equations are broadly classified into ordinary differential equations and partial differential equations. If a differential equation involves derivatives of an unknown function with respect to one independent variable, it is called an *ordinary differential equation.* If it involves partial derivatives of an unknown function of two or more independent variables, it is a *partial differential equation.*

For example,

$5\frac{dy}{dx}+2y=3x$ is an ordinary differential equation;

$\frac{\partial^2 z}{\partial x^2}+\frac{\partial z}{\partial y}=0$ is a partial differential equation.

In this chapter we study only ordinary differential equations.

AN INTRODUCTION TO DIFFERENTIAL EQUATIONS

An introduction to Differential Equation we consider briefly some important concept which we shall require for study of Differential Equation.

PFAFFIAN DIFFERENTIAL EQUATIONS

Pfaffian Differential Form is an expression

$$\sum_{i=1}^{n} F_i(x_1, x_2, \ldots x_n)\, dx_i \qquad \ldots(1)$$

where F_i are functions of some or all of the n variables x_1, x_2 ... x_n.

The equation

$$\sum_{i=1}^{n} F_i\, dx_i = 0 \qquad \ldots(2)$$

is called *Pfaffian Differential Equation.*

Theorem:

A Pfaffian Differential Equation in two variables always possesses an integrating factor.

Proof:

If x and y be two variables, Pfaffian Differential Equation is of the form

$$Mdx + Ndx = 0 \qquad \ldots(3)$$

which can be written in the form

$$\frac{dy}{dx} = f(x, y) \qquad \ldots(4)$$

where $f(x, y) = \frac{-M}{N}$

If M, N are single valued, $\frac{dy}{dx}$ is also single valued and hence solution of equation (3) consist of one-parameter family of curves in xy-plane.

$\Rightarrow$ At least in a certain region of xy-plane there exists exactly one function $\phi(x, y)$ such that

$$\phi(x, y) = c \quad \ldots(5)$$

where c is a constant, is a solution of (3)

Differentiating (5)

$$\frac{\partial \phi}{\partial x} dx + \frac{\partial \phi}{\partial y} = \mu \text{ (say)} \quad \ldots(6)$$

Comparing (3) and (6)

$$\frac{1}{M}\frac{\partial \phi}{\partial x} = \frac{1}{N}\frac{\partial \phi}{\partial y} = \mu \text{ (say)}$$

$\therefore$ On multiplying eq. (3) by μ, it takes the form

$$0 = \mu M dx + \mu N dy$$

$$= \frac{\partial \phi}{\partial x} dx + \frac{\partial \phi}{\partial y} dy$$

$$= d\phi$$

$\therefore$ μ, is an integrating factor of (3)

For three variable x, y, z the Pfaffian differential equation is of the form

$$Pdx + Qdy + Rdz = 0 \quad \ldots(7)$$

where P, Q, R are functions of x, y, z.

In vector from the eq. (7) can be written as

$$X.dR = 0 \quad \ldots(8)$$

It will be proved that eq. (8) is integrable if and only if

$$X, \text{curl } X = 0$$

The solution of eq. (7) is a relation between x, y, z (containing at the most one arbitrary constant) satisfying the differential equation.

FORMATION OF DIFFERENTIAL EQUATIONS

Given an n-parameter family, an nth order differential equation can be obtained whose solution is the given family. The technique used is to perform n successive differentiations and eliminate the n parameters $c_1, c_2, \ldots c_n$ from the system.

INTEGRAL OF A FUNCTION

Integration is an important tools of differential equation. Thus, we should have knowledge of differential method of integration. Here we describe some important method of integration.

Integral of a Function

Definition: *The integral of a function f(x) w.r.t. x, denoted* $\int f(x)\,dx$, *is defined as a function g(x) such that*

$$g'(x) = f(x). \qquad \ldots(1)$$

Thus, $\int f(x)\,dx = g(x) \Leftrightarrow g'(x) = f(x)$.

The function g(x) is called an *integral* of f(x).

It may be noted that if g(x) is an integral of f(x), then g(x) + c is also an integral of f(x), where c is an arbitrary constant, since

$$\frac{d}{dx}[g(x) + c] = \frac{d}{dx}[g(x)] + 0 = g'(x) = f(x), \text{ by (1)},$$

c is called the *constant of integration.*

Remark:

$$\frac{d}{dx}\left[\int f(x)\,dx\right] = f(x).+c$$

Let $\int f(x)\,dx = g(x)$.

Then $\frac{d}{dx}\left[\int f(x)\,dx\right] = g'(x)$, by (1).

ORDER AND DEGREE OF A DIFFERENTIAL EQUATION

The order of a differential equation is determined by the order of the highest derivative in the equation. The degree of a differential equation is determined by the highest degree of the derivatives, when the equation is rationalized.

For example,

i. $5\frac{dy}{dx}+2y=3x$ is of order 1 and degree 1

ii. $5\left(\frac{dy}{dx}\right)^2+2y=3x$ is of order 1 and degree 2

iii. $5\frac{d^2y}{dx^2}+2y=3x$ is of order 2 and degree 1

iv. $5\left(\frac{d^2y}{dx^2}\right)^2+2y=3x$ is of order 2 and degree 2

INDEPENDENCE OF CONSTANTS OF INTEGRATION

Let $y = \phi(x, c_1, c_2)$ be the solution of an equation of the second order.

The constants c_1 and c_2 are said to be independent if this solution is not reducible to the form $y = \psi\{x, f(c_1, c_2)\}$, in which there is really only one arbitrary constant $f(c_1, c_2)$.

Theorem 1:

The necessary and sufficient condition that the parameters c_1 and c_2 in $\phi(x, c_1, c_2)$ are independent is that the determinant

$$\begin{vmatrix} \dfrac{\partial\phi}{\partial c_1} & \dfrac{\partial\phi}{\partial c_2} \\ \dfrac{\partial^2\phi}{\partial c_1\partial x} & \dfrac{\partial^2\phi}{\partial c_2\partial x} \end{vmatrix}$$

be not equal to zero.

Proof:

Necessary Condition: Let us consider $\phi(x, c_1, c_2)$ can be put in the form $\psi\{x, f(c_1, c_2)\}$ then we have

$$\frac{\partial\phi}{\partial c_1} = \frac{\partial\psi}{\partial f}.\frac{\partial f}{\partial c_1}, \quad \frac{\partial\phi}{\partial c_2} = \frac{\partial\psi}{\partial f}.\frac{\partial f}{\partial c_2}$$

$$\frac{\partial^2\phi}{\partial c_1\partial x} = \frac{\partial}{\partial c_1}\left(\frac{\partial\phi}{\partial x}\right) = \frac{\partial}{\partial c_1}\left(\frac{\partial\psi}{\partial x}\right) = \frac{\partial}{\partial f}\left(\frac{\partial\psi}{\partial x}\right).\frac{\partial f}{\partial c_1} = \frac{\partial^2\psi}{\partial f\partial x}.\frac{\partial f}{\partial c_1}$$

Similarly, $$\frac{\partial^2\phi}{\partial c_2\partial x} = \frac{\partial^2\psi}{\partial f\partial x}\frac{\partial f}{\partial c_2}$$

$$\therefore \begin{vmatrix} \dfrac{\partial\phi}{\partial c_1} & \dfrac{\partial\phi}{\partial c_2} \\ \dfrac{\partial^2\phi}{\partial c_1\partial x} & \dfrac{\partial^2\phi}{\partial c_2\partial x} \end{vmatrix} = \begin{vmatrix} \dfrac{\partial\psi}{\partial f}.\dfrac{\partial f}{\partial c_1} & \dfrac{\partial\psi}{\partial f}.\dfrac{\partial f}{\partial c_2} \\ \dfrac{\partial^2\psi}{\partial f\partial x}.\dfrac{\partial f}{\partial c_1} & \dfrac{\partial^2\psi}{\partial f\partial x}.\dfrac{\partial f}{\partial c_2} \end{vmatrix}$$

$$= \frac{\partial\psi}{\partial f}.\frac{\partial f}{\partial c_1}.\frac{\partial^2\psi}{\partial f\,\partial x}.\frac{\partial f}{\partial c_2} - \frac{\partial\psi}{\partial f}.\frac{\partial f}{\partial c_2}.\frac{\partial^2\psi}{\partial f\,dx}.\frac{\partial f}{\partial c_1} = 0.$$

Hence, if in $f(x, c_1, c_2)$, c_1 and c_2 are independent then the determinant is not equal to zero.

Sufficient Condition: If the determinant be identically zero then $\phi(x, c_1, c_2)$ must be of the form $y(x, f(c_1, c_2)\}$, i.e., $f(x, c_1, c_2)$ will not vary, no matter how c_1 and c_2 are varied, so long as $f(c_1, c_2)$ is assigned some particular constant value.

Let $\dfrac{\partial \phi}{\partial c_1} = p$ and $\dfrac{\partial \phi}{\partial c_2} = q.$

$$\therefore \quad \begin{vmatrix} \dfrac{\partial \phi}{\phi c_1} & \dfrac{\partial \phi}{\partial c_2} \\ \dfrac{\partial^2 \phi}{\partial c_1 \partial x} & \dfrac{\partial^2 \phi}{\partial c_2 \partial x} \end{vmatrix} = 0 \quad \text{gives} \quad \begin{vmatrix} p & q \\ \dfrac{\partial p}{\partial x} & \dfrac{\partial q}{\partial x} \end{vmatrix} = 0$$

$$\Rightarrow \qquad p\frac{\partial q}{\partial x} - q\frac{\partial p}{\partial x} = 0. \qquad \therefore \frac{p}{q} = a(\text{constant})$$

i.e., p/q is independent of x, and hence can only involve c_1 and c_2. Let us take $p/q = L/M$,

where L, M are functions of c_1 and c_2

$$\therefore \qquad Mp = Lq$$

$\Rightarrow M\dfrac{\partial \phi}{\partial c_1} = L\dfrac{\partial \phi}{\partial c_2}$. Now from $\phi(x, c_1, c_2)$, we have

$$d\phi = \frac{\partial \phi}{\partial x}.dx + \frac{\partial \phi}{\partial c_1}.dc_1 + \frac{\partial \phi}{\partial c_2}.dc_2$$

$$= \frac{\partial \phi}{\partial x}.dx + \frac{\partial \phi}{\partial c_1}.dc_1 + \frac{M}{L}.\frac{\partial \phi}{\partial c_2}.dc_2$$

$$= \frac{\partial \phi}{\partial x}.dx + (L dc_1 + M dc_2)\frac{1}{L}.\frac{\partial \phi}{\partial c_1}.$$

But, $L\, dc_1 + M\, dc_2$ has an integrating factor μ such that $\mu L\, dc_1 + M\mu dc_2$ is a complete differential of the form $df(c_1, c_2)$.

$$\therefore \qquad d\phi = \frac{\partial \phi}{\partial x}.dx + \frac{1}{\mu L}\frac{\partial \phi}{\partial c_1} df(c_1, c_2)$$

Hence, $\phi(x, c_1, c_2)$ will not vary, no matter how c_1, c_2 are varied, provided only that they satisfy the condition

$$f(c_1, c_2) = a \text{ (constant)}.$$

Hence, the necessary and sufficient condition that c_1, c_2 are independent in $\phi(x, c_1, c_2)$ is that the above determinant be not equal to zero.

In general, the necessary and sufficient condition that the n parameters $c_1, c_2, ..., c_n$ in $f(x, c_1, c_2, ..., c_n)$ be independent is that the determinant

$$\begin{vmatrix} \frac{\partial\phi}{\partial c_1} & \frac{\partial\phi}{\partial c_2} & \cdots & \frac{\partial\phi}{\partial c_n} \\ \frac{\partial^2\phi}{\partial c_1\partial x} & \frac{\partial^2\phi}{\partial c_2\partial x} & \cdots & \frac{\partial^2\phi}{\partial c_n\partial x} \\ \cdots & \cdots & \cdots & \cdots \\ \cdots & \cdots & \cdots & \cdots \\ \frac{\partial^n\phi}{\partial c_1\partial x^{n-1}} & \frac{\partial^n\phi}{\partial c_n\partial x^{n-1}} & \cdots & \frac{\partial^n\phi}{\partial c_n\partial x^{n-1}} \end{vmatrix}$$

be not equal to zero.

Linear dependence and independence of solutions of an equation.

Let $y_1, y_2, ..., y_n$ be n solutions of the linear differential equation.

$$\frac{d^ny}{dx^n} + P_1\frac{d^{n-1}y}{dx^{n-1}} + ... + P_ny = 0,$$

where $P_1, P_2, ..., P_n$ are functions of x or constant.

These n solutions $y_1, y_2, ..., y_n$ are said to be *linearly dependent,* if there exists a set of n constants $a_1, a_2, ..., a_n$ exists then the solutions $y_1, y_2, ..., y_n$ are said to be linearly independent.

Theorem 2:

The necessary and sufficient condition that n integrals $y_1, y_2, ..., y_n$ of the linear differential equation

$$\frac{d^ny}{dx^n} + P_1\frac{d^{n-1}y}{dx^{n-1}} + ... + P_ny = 0 \qquad ...(i)$$

where $P_1, P_2, ..., P_n$ are functions of x or constants form a system of linearly independent integrals is that the determinant,

$$\begin{vmatrix} y_1 & y_2 & \cdots & y_n \\ y'_1 & y'_2 & \cdots & y'_n \\ y''_1 & y''_2 & \cdots & y''_n \\ \cdots & \cdots & \cdots & \cdots \\ \cdots & \cdots & \cdots & \cdots \\ y_1^{n-1} & y_2^{n-2} & \cdots & y_n^{n-1} \end{vmatrix}$$

does not vanish identically:

Proof:

Let us consider there is a linear relation $a_1y_1 + a_2y_2 + ... + a_ny_n = 0$, where $a_1, a_2, ..., a_n$ are constants, existing between all or any of the integrals $y_1, y_2,$

..., y_n then the integral

$$y = c_1y_1 + c_2y_2 + ... + c_ny_n, \text{ may be written as}$$

$$y = \frac{c_1}{a_1}[-a_2y_2 - a_3y_3 ... -a_ny_n] + c_2y_2 + ... + c_ny_n$$

$$\Rightarrow \quad y = \left(c_2 - c_1\frac{a_2}{a_1}\right)y_2 + \left(c_3 - c_1\frac{a_3}{a_1}\right)y_3 + ... + \left(c_n - c_1\frac{a_n}{a_1}\right)y_n,$$

which does not really contain more than (n – 1) arbitrary constants and therefore is not the general solution of (i).

Suppose that the condition holds in the case of (n – 1) functions, then is holds for n functions.

Necessary Condition: Let $y_1, y_2, ..., y_n$ are not linearly independent then there exists a relation

$$a_1y_1 + a_2y_2 + ... + a_ny_n = 0 \qquad ...(ii)$$

between them where $a_1, a_2, ..., a_n$ are constants, not all equal to zero.

$\therefore$ If $a_1 \neq 0$, then the determinant

$$R = \begin{vmatrix} y_1 & y_2 & ... & y_n \\ y'_1 & y'_2 & ... & y'_n \\ ... & ... & ... & ... \\ ... & ... & ... & ... \\ y_1^{n-1} & y_2^{n-2} & ... & y_n^{n-1} \end{vmatrix}$$

$$= \frac{1}{a_1}\begin{vmatrix} a_1y_1 + a_2y_2 + ... + a_ny_n & y_2 ... y_n \\ a_1y'_1 + a_2y'_2 + ... + a_ny'_n & y'_2 ... y'_n \\ ... \quad ... & ... \\ ... \quad ... & ... \\ a_1y_1^{n-1} + a_2y_2^{n-1} + ... + a_ny_n^{n-1} & y_2^{n-1} ... y_n^{n-1} \end{vmatrix}$$

$$= \frac{1}{a_1}\begin{vmatrix} 0 & y_2 & ... & y_n \\ 0 & y'_2 & ... & y'_n \\ ... & ... & ... & ... \\ ... & ... & ... & ... \\ 0 & y_2^{n-1} & ... & y_n^{n-1} \end{vmatrix} = 0$$

i.e., if $y_1, y_2, ..., y_n$ are not linearly independent then the determinant R will be identically equal to zero.

Sufficient Condition: If R = 0, there will be a linear relation of the form (ii) between the integrals $y_1, y_2, ..., y_n$.

Since R = 0, the determinant must be reducible to zero wherein all the elements of one column (or row) are zero. Therefore, there must be certain multiplies $\lambda_1, \lambda_2, ..., \lambda_n$ such that

$$\left.\begin{array}{lllll} \lambda_1 y_1 & +\lambda_2 y_2 & +...+\lambda_n y_n & = & 0 \\ \lambda_1 y'_1 & +\lambda_2 y'_2 & +...+\lambda_n y_n & = & 0 \\ ... & ... & ... & ... & ... \\ ... & ... & ... & ... & ... \\ \lambda_1 y_1^{n-1} & +\lambda_2 y_2^{n-1} & +...+\lambda_n y_n^{n-1} & = & 0 \end{array}\right] \quad ...(iii)$$

Differentiating each of these equation and subtracting each from the next following, we have

$$\left.\begin{array}{lllll} \lambda'_1 y_1 & +\lambda'_2 y_2 & +...+\lambda'_n y_n & = & 0 \\ \lambda'_1 y'_1 & +\lambda'_2 y'_2 & +...+\lambda'_n y'_n & = & 0 \\ ... & ... & ... & ... & ... \\ ... & ... & ... & ... & ... \\ \lambda'_1 y_1^{n-2} & +\lambda'_2 y_2^{n-2} & +...+\lambda'_n y_n^{n-2} & = & 0 \end{array}\right] \quad ...(iv)$$

If one of the determinants

$$\begin{bmatrix} y_1 & y_2 & ...y_n \\ y'_1 & y'_2 & ...y'_n \\ ... & ... & ... \\ ... & ... & ... \\ y_1^{n-2} & y_2^{n-2} & ...y_n^{n-2} \end{bmatrix}$$

vanishes, say the one formed by omitting the rth column, then there is a relation. $c_1 y_1 + c_2 y_2 + ... + c_{r-1} y_{r-1} + c_{r+1} y_{r+1} + ... + c_n y_n = 0$

But if no one of these determinants vanishes, then from (iv) and the first (n – 1) equations in (iii), we have

$$\frac{\lambda'_1}{\lambda_1} = \frac{\lambda'_2}{\lambda_2} = \dots = \frac{\lambda'_n}{\lambda_n} = \mu \text{ (say)} \quad \therefore \lambda'_1 = \lambda_1\mu \quad \Rightarrow \lambda_1 = a_1 e^{\int \mu dx}$$

Similarly, $\lambda_2 = a_2 e^{\int \mu dx}, \dots, \lambda_n = a_n e^{\int \mu dx}$

$a_1, a_2, \dots, a_n$ being constants.

Substituting these in the first of the equations (iii), we have

$$a_1 y_1 + a_2 y_2 + \dots + a_n y_n = 0.$$

Hence, if the condition holds for n – 1 functions, it also holds for n. But as in 1.5 it can be shown that it hold for n = 2 and hence it holds for n = 3, and so on for any number.

Hence, the necessary and sufficient condition that $y_1, y_2, \dots, y_n$ form a system of linearly independent integrals, is that the determinant R does not vanish identically.

Remarks:

1. *The system of linearly independent integrals in sometimes called a Fundamental System of Integrals.*
2. *The determinant R is also known as the* **Wronskian**. *The Wronskian is named after Hoene Wronski, one of the early writers on determinants.*

Wronskian: *Def. The determinant.*

$$W(x) \begin{vmatrix} y_1 & y_2 & \dots & y_n \\ y'_1 & y'_2 & \dots & y'_n \\ y''_1 & y''_2 & \dots & y''_n \\ \dots & \dots & \dots & \dots \\ \dots & \dots & \dots & \dots \\ y_n^{n-1} & y_2^{n-1} & \dots & y_n^{n-1} \end{vmatrix}$$

is called the **Wronskian** of the set of function $y_1, y_2, y_3, \dots, y_n$.

Fundamental Sets of Solutions: A set of linearly independent solutions is called a fundamental set of solutions.

SOLUTION OF ORDINARY DIFFERENTIAL EQUATION

A solution of an ordinary differential equation is a functional relationship, explicit or implicit, between the variables involved, not containing derivatives or differentials, which satisfies the differential equation. The *general solution* of an nth order differential equation is a solution containing n independent arbitrary constants of integration. A *particular solution* of a differential equation is a solution obtained from the general solution by giving particular values to the arbitrary constants. For the differential equation given by equation (1) at the beginning has general solution $y = x^2 + c$; while, $y = x^2$, or $y = x^2 + 1$ is a particular solution. Some differential equations also have solutions that cannot be obtained as particular solutions. These solutions are called *singular solutions*. As in the case of integration, the constant of integration arose while solving a differential equation can be determined using initial or boundary conditions. The existence and uniqueness of the solution of a differential equation is established by the *Fundamental Existence Theorem*. The theorem states that:

Consider the differential equation $\frac{dy}{dx} = f(x, y)$, where $f(x, y)$ is continuous and its derivatives with respect to y are continuous in a region R of the xy-plane. Then for each point (x_0, y_0) in R,

(*i*) there exists a solution $f(x)$ of the differential equation, and

(*ii*) there is one and only one solution such that $f(x_0) = y_0$.

Example:

Show that $y = A \cos 2x + B \sin 2x$, where A and B are arbitrary constants, is the general solution of the differential equation:

$$\frac{d^2y}{dx^2} + 4y = 0$$

Solution:

Given that $y = A \cos 2x + B \sin 2x$

Then, $\frac{dy}{dx} = (-2) A \sin 2x + 2B \cos 2x$

and $\frac{d^2y}{dx^2} = -4A \cos 2x - 4B \sin 2x$

$$\frac{d^2y}{dx^2} + 4y = -4A \cos 2x - 4B \sin 2x + 4 (A \cos 2x + B \sin 2x) = 0$$

Hence, $y = A \cos 2x + B \sin 2x$ is the general solution of $\frac{d^2y}{dx^2} + 4y = 0.$

FORMATION OF DIFFERENTIAL EQUATIONS

Differential equations are formed by eliminating the arbitrary constants involved in a relation connecting the variables using differentiation.

Example:

Find the differential equation of all straight lines passing through the origin.

Solution:

Equation of any straight line passing through the origin is of the form:

$$y = mx \qquad \text{...(1)}$$

m being the parameter.

Differentiating (1) with respect to x, we get

$$\frac{dy}{dx} = m \qquad \text{...(2)}$$

Eliminating m between (1) and (2), we get

$$m = \frac{y}{x} \qquad \text{[from (1)]}$$

$$\frac{dy}{dx} = \frac{y}{x} \qquad \text{[Substituting in (2)]}$$

∴ The required differential equation is:

$$x\frac{dy}{dx} - y = 0$$

DIFFERENTIAL EQUATION

Definition: An equation which involve independent and dependent variables and the differential co-efficient derivatives of the dependent variable with respect to the independent variable of variables are called *Differential Equations.*

Examples:

$$(4x+5y)\frac{dy}{dx}+(3x+2y+5) = 0 \qquad \text{...(i)}$$

$$\frac{d^2y}{dx} = a^2x = 0 \qquad \text{...(i)}$$

$$\left(\frac{dy}{dx}\right)^2 = 4\left(\frac{dy}{dx}\right) - 2y^2 + 4x \qquad ...(iii)$$

$$\frac{\partial z}{\partial x} + \frac{\partial z}{\partial y} = kz \qquad ...(iv)$$

Ordinary Differential Equations: Equations which involve only one independent variable are called *Ordinary Differential Equations.*

Partial Differential Equations: Which involve partial differential coefficients with respect to more than one independent variable are called *Partial Differential Equations.*

Order and Degree of a Differential Equation: The order of differential equation is the order of the highest derivative involved in the differential equation. Equation (i) is of second order while (ii) is of first order.

The degree of a differential equation is the power of the highest order derivative involved in the equation, when the equation has been made rational and integral as far as the derivatives are concerned.

Example:

To determine the degree of the equation

$$\frac{d^2y}{dx^2} = k\left[1 + \left(\frac{dy}{dx}\right)^2\right]^{5/2}$$

Solution:

Firstly, we square the equation to rationalise it and see that the power of highest derivative in the equation becomes two. So the degree of the above differential equation is two.

Solution of the Differential Equation: By a solution of a differential equation we mean any functional relation among the variables free of the derivatives, which among with the derivative obtained from them satisfies the given differential equation.

LINEAR DIFFERENTIAL EQUATIONS OF SECOND ORDER WITH CONSTANT COEFFICIENTS

The general form of a linear differential equation of the second order with constant coefficients is:

$$a_0 \frac{d^2y}{dx^2} + a_1 \frac{dy}{dx} + a_2 y = f(x) \qquad ...(1)$$

For solving this, first we set .

i.e., $$a_0 \frac{d^2y}{dx^2} + a_1 \frac{dy}{dx} + a_2 y = 0 \qquad \ldots(2)$$

The equation (2) is called homogeneous equation associated with (1), and its solution is called *complementary function(CF)* of (1). Since the general solution of a differential equation of the second order contains 2 arbitrary constants, if $f_1(x)$ and $f_2(x)$ are any two independent solutions of (1), then, $Af_1(x) + Bf_2(x)$, where and are arbitrary constants, which is also a solution of (2), will be the general solution of (1). Again if *PI* is any *particular integral* of (1), then, it can be shown that:

$$Af_1(x) + Bf_2(x) + PI$$

is the general solution of (1).

Hence the general solution of (1) is:

$$y = \text{Complementary function} + \text{Particular integral} = CF + PI$$

Now for solving (2), we suppose e^{mx} as a trial solution of (2).

Then we have, $e^{mx}(a_0m^2 + a_1m + a_2) = 0$

This means, e^{mx} will be a solution of (2) if m is a root of the algebraic equation,

$$a_0m^2 + a_1m + a_2 = 0 \qquad \ldots(3)$$

Equation (3) is called *auxiliary equation* or characteristic equation.

Now three cases arise.

Case-1 (The two roots are real and distinct)

Let m_1 and m_2 are the roots of equation (3), where m_1 and m_2 are real and distinct. Then e^{m_1x} and e^{m_2x} are different and linearly independent. Hence, the general solution of (2) is:

$$Ae^{m_1x} + Be^{m_2x}$$

Case-2 (The two roots are equal)

Let m be the equal root of equation (3). Then e^{m_1x} and e^{m_2x} are different and linearly independent. Hence, the general solution of (2) reduces to $C\,e^{mx}$, where C is an arbitrary constant [$C = A + B$]. Since it contains only one arbitrary constant, it cannot be taken as the general solution.

By proceeding further we get the general solution of equation (3) as:

$$(A + Bx)e^{mx}$$

<u>Case-3</u> (The two roots are complex)

Let the two roots of equation (3) are $\alpha + i\beta$ and $\alpha - i\beta$.

$\therefore$ The general solution

$$= A_1 e^{(\alpha+i\beta)x} + B_1 e^{(\alpha-i\beta)x}$$

$$= e^{\alpha x}\left[A_1 e^{i\beta x} + B_1 e^{-i\beta x}\right]$$

$$= e^{\alpha x}\left[A_1 e^{i\beta x} + B_1 e^{-i\beta x}\right]$$

$$= e^{\alpha x}\left[A_1(\cos\beta x + i\sin\beta x) + B_1(\cos\beta x - i\sin\beta x)\right]$$

$$= e^{\alpha x}\left[A\cos\beta x + B\sin\beta x\right];$$

where $A = A_1 + B_1$ and $B = i(A_1 - B_1)$

Example:

Solve: $\dfrac{d^2y}{dx^2} + 5\dfrac{dy}{dx} + 6y = 0$

Solution:

Since the given equation is homogeneous, it has only complementary solution as general solution.

Here the auxiliary equation is $m^2 + 5m + 6 = 0$

Roots are $m_1 = -2$ and $m_2 = -3$

Complimentary Function, $CF = Ae^{-2x} + Be^{-3x}$

$\therefore$ General Solution is: $y = Ae^{-2x} + Be^{-3x}$

Note: We can check this solution by substituting the value in the given equation.

Put $y = Ae^{-2x} + Be^{-3x}$ in the given equation.

We get $\dfrac{dy}{dx} = \left(-2Ae^{-2x} - 3Be^{-3x}\right)$, and $\dfrac{d^2y}{dx^2} = 4Ae^{-2x} + 9Be^{-3x}$

Hence the given equation becomes,

$$4Ae^{-2x} + 9Be^{-3x} + 5\left(-2Ae^{-2x} - 3Be^{-3x}\right) + 6\left(Ae^{-2x} + Be^{-3x}\right) = 0$$

Hence the result.

Method of Finding Particular Integral

Consider equation (1) above.

i.e., $$a_0 \frac{d^2y}{dx^2} + a_1 \frac{dy}{dx} + a_2 y = f(x)$$

Writing , the equation can be put in the form:

$$a_0 D^2 y + a_1 Dy + a_2 y = f(x)$$

i.e., $$(a_0 D^2 + a_1 D + a_2)y = f(x)$$

i.e., $$\varphi(D) = f(x), \text{ where } \varphi(D) \equiv a_0 D^2 + a_1 D + a_2$$

<u>Case-i</u> When $f(x) = e^{ax}$

We have
$$\begin{aligned} \varphi(D)e^{ax} &= (a_0 D^2 + a_1 D + a_2)e^{ax} \\ &= a_0 D^2(e^{ax}) + a_1 D(e^{ax}) + a_2 e^{ax} \\ &= a_0 a^2 e^{ax} + a_1 a e^{ax} + a_2 e^{ax} \\ &= (a_0 a^2 + a_1 a + a_2)e^{ax} = \varphi(a)e^{ax} \end{aligned} \qquad \ldots(4)$$

Consider $\frac{1}{\varphi(D)}$ as the inverse operator of $\varphi(D)$.

Now applying the operator on both sides of (4),

$$\frac{1}{\varphi(D)}\varphi(D)e^{ax} = \frac{1}{\varphi(D)}\varphi(a)e^{ax}$$

i.e., $$e^{ax} = \varphi(a)\frac{1}{\varphi(D)}e^{ax}$$

i.e., $$\frac{1}{\varphi(D)}e^{ax} = \frac{1}{\varphi(a)}e^{ax}, \text{ provided } \varphi(a) \neq 0$$

Remark 1:

This method fails when $\varphi(a) = 0$.

If a is a non-repeated root of $\varphi(D)$, then we write

$$\varphi(D) = (D - a)\psi(D),, \text{ where } \psi(D) \neq 0$$

Then we get,

$$\frac{1}{\varphi(D)}e^{ax} = \frac{x}{\varphi(a)}e^{ax}$$

If a is repeated root of $\varphi(D)$, then we get,

$$\frac{1}{\varphi(D)}e^{ax} = \frac{x^2}{2}e^{ax}$$

<u>Case-ii</u> When $f(x) = \sin ax$ or $\cos ax$

We have $\varphi(D^2) \equiv a_0D^4 + a_1D^2 + a_2$

$$\begin{aligned}\therefore \quad \varphi(D^2)\sin ax &= (a_0D^4 + a_1D^2 + a_2)\sin ax \\ &= a_0(-a^2)^2 \sin ax + a_1(-a^2)\sin ax + a_2 \sin ax \\ &= \varphi(-a^2)\sin ax\end{aligned}$$

Operating by $\frac{1}{\varphi(D^2)}$ on both sides,

$$\frac{1}{\varphi(D^2)}\varphi(D^2)\sin ax = \frac{1}{\varphi(D^2)}\varphi(-a^2)\sin ax$$

i.e., $$\sin ax = \varphi(-a^2)\frac{1}{\varphi(D^2)}\sin ax$$

$$\therefore \quad \frac{1}{\varphi(D^2)}\sin ax = \frac{1}{\varphi(-a^2)}\sin ax; \text{ provided } \varphi(-a^2) \neq 0$$

Remark 2:

This method fails when $\varphi(-a^2) = 0$. Then we proceed as follows:

$$\begin{aligned}\frac{1}{D^2 + a^2}\sin ax &= \frac{1}{D^2 + a^2}\left(\text{Imaginary Part of } e^{iax}\right) \\ &= \text{Imaginary Part of } \frac{1}{D^2 + a^2}\left(e^{iax}\right) \\ &= \text{Imaginary Part of } \frac{1}{(D - ia)(D + ia)}\left(e^{iax}\right)\end{aligned}$$

$$= \text{Imaginary Part of } \frac{1}{(D-ia)}\frac{1}{2ia}\left(e^{iax}\right), \text{ by case}$$

$$= \text{Imaginary Part of } \frac{1}{2ia(D-ia)}\left(e^{iax}\right)$$

$$= \text{Imaginary Part of } \frac{1}{2ia}xe^{iax}, \text{ by the Remark-1}$$

$$= \text{Imaginary Part of } \frac{-ix}{2a}(\cos ax + i\sin ax)$$

$$= \frac{-x\cos ax}{2a}$$

Same method may be used for $\cos ax$ also, in which case we consider Real Part, instead of Imaginary Part.

<u>Case-iii</u> When $f(x) = x^m$, m being a positive integer

Here we have to evaluate $\frac{1}{\varphi(D)}x^m$. First expand $\frac{1}{\varphi(D)} = (\varphi(D))^{-1}$ in ascending integral powers of D, and then let each term of the expansion operate on x^m.

Example:

Solve: $\frac{d^2y}{dx^2} - 13\frac{dy}{dx} + 12y = e^{-2x}$

Solution:

The auxiliary equation is $m^2 - 13m + 12 = 0$

Roots are $m_1 = 1$ and $m_2 = 12$

Complimentary Function, $CF = Ae^x + Be^{12x}$

Particular Integral,

$$PI = \frac{1}{D^2 - 13D + 12}e^{-2x}$$

$$= \frac{1}{(-2)^2 - 13(-2) + 12}e^{-2x} = \frac{1}{42}e^{-2x}$$

$\therefore$ General Solution is: $y = Ae^{x} + Be^{12x} + \frac{1}{42}e^{-2x}$

GENERAL AND PARTICULAR SOLUTION

General solution is that solution of a differential equation which contain the number of arbitrary constants equal to the order of the differential equation. It is also called *Complete Integral.*

A particular solution of a differential equation is a solution obtain from the general solution by giving particular values to the arbitrary constants.

THE COMPLETE SOLUTION OF A DIFFERENTIAL EQUATION OF THE NTH ORDER CONTAINS N ARBITRARY INDEPENDENT CONSTANTS

Let y', y'', ... denote the first, second, ..., order, derivatives of y with respect to, x and y(0), y'(0), y''(0), ... denote the values of y, y', y'', ... when x = 0.

Given as consider an equation of the first order. Let the equation be

$$f(y', y, x) = 0 \qquad \ldots(1)$$

Let $\quad y = c + c_1x + c_2x^2 + \ldots \qquad \ldots(2)$

be the solution of (1), when *expended in ascending powers of x.*

From Maclaurin's Theorem, we have

$$y(x) = y(0) + y'(0).x + y''(0)\,.\,\frac{x^2}{2!} + Y'''(0)\,.\,\frac{x^3}{3!} + \ldots \qquad \ldots(3)$$

Comparing (2) and (3), we have

$$c = y(0),\ c_1 = y'(0),\ c_2 = \frac{y''(0)}{2!},\ C_3 = \frac{y'''(0)}{3!},\ \ldots$$

Now, c = y (0) *cannot be expressed in terms of anything known* or determinate.

Since (1) holds true for all values of x, for x = 0 we obtain

$$f\{y'(0), y(0), 0\} = 0$$

$$\Rightarrow \qquad f(c_1, c, 0) = 0$$

Which determines c_1 in terms of c.

We solve equation (1) for

Thus, $\quad y' = \phi(y, x)$

$$\therefore\ y'' = \frac{\partial\phi}{\partial x} + \frac{\partial\phi}{\partial y}y'$$

Then forx = 0, we get

$$y''(0) = \left(\frac{\partial\phi}{\partial x}\right)_{x=0} + \left(\frac{\partial\phi}{\partial y}\right)_{x=0} .y'(0) \Rightarrow 2!\, c_2 = \left(\frac{\partial\phi}{\partial x}\right)_{x=0} + \left(\frac{\partial\phi}{\partial y}\right)_{x=0} c_1.$$

which determines c_2 in terms of c_1. But c_1 is found in terms of c, therefore, c_2 may also be found in terms of c.

Proceeding similarly by differentiating and substituting x = 0, we shall find that c_3, c_4 .. may be expressed in terms of c. Therefore all the constants except c are determined. Thus, the differential equation of the first order has one arbitrary constant in its general solution.

In the next place, let us consider an equation of the second order as F(y", y', y, x) = 0 ...(4)

Let the solution of (iv) be $y = c + c_1x + c_2x^2 + \ldots$

As before, we know that

$$c = y(0),\ c_1 = y'(0),\ c_2 = \frac{1}{2!}y''(0),\ c_3 = \frac{1}{3!}.y'''(0), \ldots$$

Now equation (4), may be solve for y". Thus, we have

$$y'' = \psi\ (y, y, x) \qquad \ldots(5)$$

Then for x = 0, we have

$$y''(0) = \psi\ \{y'(0), y(0), 0\}$$

$$\Rightarrow \quad 2!\ c_2 = \psi\ (c_1, c, 0)$$

which determines c_2 in terms of c and c_1.

Again differentiating (5), we get

$$y'''(x) = \frac{\partial\psi}{\partial x} + \frac{\partial\psi}{\partial y}y' + \frac{\partial\psi}{\partial y'}y''$$

Then for x = 0, $y'''(0) = \left(\frac{\partial\psi}{\partial x}\right)_{x=0} + \left(\frac{\partial\psi}{\partial y}\right)_{x=0} y'(0) + \left(\frac{\partial\psi}{\partial y'}\right)_{x=0} y''(0)$

$$\Rightarrow \quad 3!\ c_3 = \left(\frac{\partial\psi}{\partial x}\right)_{x=0} + \left(\frac{\partial\psi}{\partial y}\right)_{x=0} .c_1 + \left(\frac{\partial\psi}{\partial y'}\right)_{x=0} 2!c_2$$

which determines c_3 in terms of c_1 and c_2. But c_2 is already found in terms of c and c_1, therefore, c_3 may also be found in terms of c and c_1.

Similarly, in this way the, values of all other co-efficients c_4, c_5 ...can be obtained in terms of c and c_1. But it will not be possible to obtain any information about c and c_1. Thus, the solution of an equation of second order contains two arbitrary constants.

Similarly, for an equation of nth order we can show that its solution contains n arbitrary constants.

Theorem:

If $y = y_1, y = y_2, \ldots, y = y_n$ are solutions of an ordinary linear differential equation of nth order of the form.

$$\frac{d^n y}{dx^n} + P_1 \frac{d^{n-1} y}{dx^{n-1}} + P_2 \frac{d^{n-2} y}{dx^{n-2}} + \ldots + P_{n-1} \frac{dy}{dx} + P_n y = 0, \quad \ldots (i)$$

where $P_1, P_2, \ldots,$ are functions of x or constants, then

$$y = c_1 y_1 = c_2 y_2 + \ldots + c_n y_n,$$

where $c_1, c_2, \ldots c_n$ are arbitrary constants, is also a solution of the equation.

Proof:

Let us consider $y = c_1 y_1 + c_2 y_2 + \ldots + c_n y_n$ ...(ii)

is the solution of equation (i), then it must satisfy (i)

$$\therefore \quad \frac{d^n}{dx^n}(c_1 y_1 + c_2 y_2 + \ldots + c_n y_n) + P_1 \frac{d^{n-1}}{dx^{n-1}}(c_1 y_1 + c_2 y_2 + \ldots + c_n y_n)$$

$$+ \ldots + P_{n-1} \frac{d}{dx}(c_1 y_1 + c_2 y_2 + \ldots + c_n y_n)$$

$$+ P_n (c_1 y_1 + c_2 y_2 + \ldots + c_n y_n)$$

$$= c_1 \left[\frac{d^n y_1}{dx^n} + P_1 \frac{d^{n-1} y_1}{dx^{n-1}} + \ldots + P_n y_1\right] + c_2 \left[\frac{d^n y_2}{dx^n} + P_1 \frac{d^{n-1} y_2}{dx^{n-1}} + \ldots + P_n y_2\right]$$

... ...

... ...

$$+ c_n \left[\frac{d^n y_n}{dx^n} + P_1 \frac{d^{n-1} y_n}{dx^{n-1}} + \ldots + P_n y_n\right] = 0$$

Since $y = y_1, y = y_2 \ldots, y = y_n$ are solution of equation (i) therefore each expression in the brackets is zero. Hence (ii) is the solution of (i).

FIRST ORDER DIFFERENTIAL EQUATIONS

The general form of a first degree (linear) ordinary differential equation is:

$$a_0(x)\frac{d^n y}{dx^n} + a_1(x)\frac{d^{n-1} y}{dx^{n-1}} + \ldots + a_{n-1}(x)\frac{dy}{dx} + a_n(x) y = f(x)$$

Clearly this is of order *n*, for $a_0(x) \neq 0$. If the coefficients of *y* and its derivatives, i.e., if $a_0(x), a_1(x), ..., a_n(x)$ are all constants, then it is called a linear differential equation with constant coefficients. In addition, if $f(x) = 0$, the differential equation is called a homogeneous linear differential equation with constant coefficients. Some methods for solving linear differential equations of the first order are discussed below.

Variable Separable Form

The differential equation $\frac{dy}{dx} = f(x, y)$ is said to be in variable separable form, if it can be put in the form:

$$M\,dx + N\,dy = 0 \qquad ...(1)$$

with M a function of x only and N a function of y only. Its solution is obtained by integrating (1) term by term.

i.e., $\int M dx + \int N dy = c$; where c is an arbitrary constant.

Illustration – 13.3

Solve : $\frac{dy}{dx} + \frac{x}{y} = 0$

Solution:

Given that $\frac{dy}{dx} + \frac{x}{y} = 0$

i.e., $\frac{dy}{dx} = -\frac{x}{y}$

then, $y\,dy = -x\,dx$

or, $x\,dx + y\,dy = 0 \qquad ...(1)$

Now, (1) is in the variable separable form.

[note that (1) is in the form M dx + N dy = 0, with M = x and N = y]

Integrating (1) we get,

$\int x dx + \int y dy$ = c ; where c is an arbitrary constant.

i.e., $\frac{x^2}{2} + \frac{y^2}{2} = \frac{a^2}{2}$, where $\frac{a^2}{2} = c$

i.e., $x^2 + y^2 = a^2$ is the required solution.

[The solution gives a family of circles, for different values of a].

Homogeneous Form

A differential equation of the form:

$$\frac{dy}{dx} = \frac{f(x,y)}{\varphi(x,y)}$$

where $f(x,y)$ and $\varphi(x,y)$ are homogeneous functions of x an y of the same degree are called homogeneous differential equation. If $f(x,y)$ and $\varphi(x,y)$ are homogeneous functions of x and y of the same degree n, then we can write them as:

$$f(x,y) = x^n f_1\left(\frac{y}{x}\right) \text{ and } \varphi(x,y) = x^n \varphi_1\left(\frac{y}{x}\right),$$

so that, $$\frac{dy}{dx} = \frac{x^n f_1\left(\frac{y}{x}\right)}{x^n \varphi_1\left(\frac{y}{x}\right)} = \frac{f_1\left(\frac{y}{x}\right)}{\varphi_1\left(\frac{y}{x}\right)} \qquad \ldots(2)$$

In (2), we put $y = vx$, so that the dependent variable y is changed to another variable v. This substitution makes the equation variable separable form, which can be solved.

Example:

Solve: $\dfrac{dy}{dx} + \dfrac{x-2y}{2x-y} = 0$

Solution:

Given that $$\frac{dy}{dx} + \frac{x-2y}{2x-y} = 0 \qquad \ldots(A)$$

It is homogeneous of degree 1.

Put $y = vx$ in (A). This gives, $\dfrac{dy}{dx} = v + x\dfrac{dv}{dx}$

Substituting these values in (A), we get,

$$v + x\frac{dv}{dx} + \frac{x-2vx}{2x-vx} = 0$$

i.e., $$v + x\frac{dv}{dx} + \frac{1-2v}{2-v} = 0$$

i.e., $$x\frac{dv}{dx} + \frac{1-v^2}{2-v} = 0$$

i.e., $$\frac{(2-v)dv}{1-v^2} + \frac{dx}{x} = 0 \qquad \text{...(B)}$$

Equation (B) is in variable separable form.

$\therefore$ The solution is: $$\int \frac{(2-v)dv}{1-v^2} + \int \frac{dx}{x} = c \qquad \text{...(C)}$$

i.e., $$\frac{1}{2}\int \left[\frac{3}{1+v} + \frac{1}{1-v}\right] dv + \int \frac{dx}{x} = c \quad \text{[using partial fraction method]}$$

i.e., $$\frac{1}{2}[3\log(1+v) - \log(1-v)] + \log x = c$$

i.e., $$3\log(1+v) - \log(1-v) + 2\log x = 2c$$

i.e., $$\log \frac{(1+v)^3 x^2}{1-v} = \log a \text{ ; where } 2c = \log a$$

i.e., $$\frac{(1+v)^3}{1-v} = x^2 a \qquad \text{...(D)}$$

Now put $v = \frac{y}{x}$. Then (D) becomes, $(x+y)^3 = a(x-y)$

EXACT DIFFERENTIAL EQUATIONS

A differential equation is said to be exact if it can be derived from its primitive directly from differentiation. A necessary and sufficient condition for a differential equation of the form $P(x,y)dx + Q(x,y)dy = 0$ to be exact is that:

$$\frac{\partial P}{\partial y} = \frac{\partial Q}{\partial x}$$

The solution may be written as the sum of two integrals; first is the integral of *P* with respect to *x* regarding *y* constant and second is the integral with respect to *y* of those terms in *Q* which do not involve *x*. Or, integrate *P* with respect to *x* regarding *y* constant and then integrate *Q* with respect to y regarding x constant, and then omit terms in the second integral which are already present in the first integral.

Example:

Solve : $(x^2 + y)dx + (y^2 + x)dy = 0$

Solution:

It is in the form $P(x, y)dx + Q(x, y)dy = 0$

Where P = $x^2 + y$, Q = $y^2 + x$

$$\frac{\partial P}{\partial y} = 1, \frac{\partial Q}{\partial x} = 1$$

Hence the given differential equation is exact.

$$\text{I integral} = \int (x^2 + y)dx$$

$$\frac{x^3}{3} + xy + c_1 = \text{[treating } y \text{ constant]}$$

$$\text{II integral} = \int (y^2 + x)dy$$

$$\frac{y^3}{3} + c_2 = \text{[treating } x \text{ constant and omitting the terms which are already present in I integral]}$$

∴ The required solution is:

$$\frac{x^3}{3} + xy + c_1 + \frac{y^3}{3} + c_2 = 0$$

i.e., $x^3 + y^3 + 3xy + 3(c_1 + c_2) = 0$

i.e., $x^3 + y^3 + 3xy + c = 0$; where $c = 3(c_1 + c_2)$

LINEAR DIFFERENTIAL EQUATIONS OF THE FIRST ORDER

We had seen that a linear differential equation is that in which the dependent variable and its differential coefficients occur only in the first degree and are not multiplied together. The standard form of a linear differential equation of the first order is:

$$\frac{dy}{dx} + Py = Q,$$

where P and Q are functions of x or constants. To solve this, multiply both sides of the equation by $e^{\int P dx}$, which is called the *integrating factor* (I.F). The solution is then written as:

$$ye^{\int P dx} = \int\left(e^{\int P dx}\right) Q dx + c$$

While solving a linear differential equation of the first order using this method, it is customary to have the coefficient of $\frac{dy}{dx}$ unity.

Example:

Solve: $\frac{dy}{dx} + \frac{1}{x} y = x$

Solution:

Given that $\frac{dy}{dx} + \frac{1}{x} y = x$

It is in the linear form $\frac{dy}{dx} + Py = Q$; where $P = \frac{1}{x}$, and $Q = x$.

$$\int P dx = \int \frac{1}{x} dx = \log x$$

$$\text{I.F} = e^{\int P dx} = e^{\log x} = x \qquad [\because e^{\log x} = x]$$

Solution is :

$$ye^{\int P dx} = \int\left(e^{\int P dx}\right) Q dx + c_1$$

i.e., $\quad yx = \int x.x \, dx + c_1$

i.e., $\quad yx = \int x^2 dx + c_1$

i.e., $\quad yx = \frac{x^3}{3} + c_1;$

or $\quad x^3 - 3xy = c;$ where $c = -3c_1$

SOLVED EXAMPLES

Example 1:

Show that

$x^2z + y^3 = Cx$, *(C is arbitrary constant) is a solution of the equation*

$(x^2z - y^3)\,dx + 3xy^2dy + x^3dz = 0$ ...(1)

Solution:

The relation is

$$\frac{x^2z + y^3}{x} = c$$

$$\therefore \quad d\left\{\frac{x^2z + y^3}{x}\right\} = 0$$

$$\Rightarrow \quad x.d(x^2z + y^3) - (x^2z + y^3)\,dx = 0$$

$$\Rightarrow \quad x\{2xzdx + x^2dz + 3y^2dy\} - (x^2z + y^3)\,dx = 0$$

i.e. $\quad (x^2z - y^3)\,dx + 3xy^2dy + x^3dz = 0$

which is given differential equation

$\therefore$ Given relation is a solution of equation (1).

Example 2:

Solve the following differential equations:

i. $e^y dx + (xe^y + 2y)dy = 0$

ii. $(2x + 3y + 1)dx + (3x - 2y + 1)dy = 0$

iii. $(xy^2)dx + (x^2y)dy = 0$

iv. $(\cos x - x\cos y)\dfrac{dy}{dx} = \sin y + y\sin x$

v. $(x^2 - 4xy - 2y^2)dx + (y^2 - 4xy - 2x^2)dy = 0$

Solution:

i. Given that $e^y dx + (xe^y + 2y)dy = 0$

It is in the form $P(x, y)dx + Q(x, y)dy = 0$; where

$$P = e^y,\ Q = xe^y + 2y$$

$$\frac{\partial P}{\partial y} = e^y;\quad \frac{\partial Q}{\partial x} = e^y$$

Hence the given differential equation is exact.

I integral $= \int e^y dx = xe^y + c_1$ [treating y constant]

II integral $= \int (xe^y + 2y) dy$

$= 2\frac{y^2}{2} + c_2$ [treating x constant and omitting the terms which are already present in I integral]

$= y^2 + c_2$

$\therefore$ The required solution is: $xe^y + c_1 + y^2 + c_2 = 0$

i.e., $xe^y + y^2 + c = 0$; where $c = c_1 + c_2$

ii. Given that $(2x+3y+1)dx + (3x-2y+1)dy = 0$

It is in the form $P(x,y)dx + Q(x,y)dy = 0$; where

$$P = 2x+3y+1, \; Q = 3x-2y+1$$

$$\frac{\partial P}{\partial y} = 3; \; \frac{\partial Q}{\partial x} = 3$$

Hence the given differential equation is exact.

I integral $= \int (2x+3y+1)dx = 2\frac{x^2}{2} + 3xy + c_1$ [treating y constant]

$= x^2 + 3xy + c_1$

II integral $= \int (3x-2y+1) dy = 2\frac{y^2}{2} + y + c_2$

[treating x constant and omitting the terms which are already present in first integral]

$= y^2 + y + c_2$

$\therefore$ The required solution is: $x^2 + 3xy + c_1 + y^2 + y + c_2 = 0$

i.e., $x^2 + y^2 + 3xy + y + c = 0$; where $c = c_1 + c_2$

iii. Given that $(xy^2)dx + (x^2y)dy = 0$

It is in the form $P(x,y)dx + Q(x,y)dy = 0$; where

$$P = xy^2, \; Q = x^2y$$

$$\frac{\partial P}{\partial y} = 2xy; \; \frac{\partial Q}{\partial x} = 2xy;$$

Hence the given differential equation is exact.

$$\text{I integral} = \int (xy^2)dx = \frac{x^2}{2}y^2 + c_1 \text{ [treating } y \text{ constant]}$$

$$\text{II integral} = \int (x^2y)\,dy = c_2$$

[treating x constant and omitting the terms which are already present in I[st] integral]

$\therefore$ The required solution is: $\frac{x^2}{2}y^2 + c_1 + c_2 = 0$

i.e., $x^2y^2 + c = 0$; where $c = 2(c_1 + c_2)$

iv. Given that $(\cos x - x\cos y)\frac{dy}{dx} = \sin y + y\sin x$

i.e., $(\sin y + y\sin x)dx - (\cos x - x\cos y)dy = 0$

It is in the form $P(x, y)dx + Q(x, y)dy = 0$; where

$$P = \sin y + y\sin x, \; Q = x\cos y - \cos x$$

$$\frac{\partial P}{\partial y} = \cos y + \sin x; \; \frac{\partial Q}{\partial x} = \cos y + \sin x$$

Hence the given differential equation is exact.

I integral $= \int(\sin y + y\sin x)dx = x\sin y - y\cos x + c_1$ [treating y constant]

II integral $= \int(x\cos y - \cos x)dy = c_2$

[treating x constant and omitting the terms which are already present in the I[st] integral]

$\therefore$ The required solution is: $x\sin y - y\cos x + c_1 + c_2 = 0$

i.e., $x\sin y - y\cos x + c = 0$; where $c = c_1 + c_2$

v. Given that $(x^2 - 4xy - 2y^2)dx + (y^2 - 4xy - 2x^2)dy = 0$

It is in the form $P(x, y)dx + Q(x, y)dy = 0$; where

$$P = x^2 - 4xy - 2y^2,\ Q = y^2 - 4xy - 2x^2$$

$$\frac{\partial P}{\partial y} = -4x - 4y;\ \frac{\partial Q}{\partial x} = -4x - 4y$$

Hence the given differential equation is exact.

I integral = $\int (x^2 - 4xy - 2y^2)dx = \frac{x^3}{3} - 2x^2y - 2xy^2 + c_1$

[treating y constant]

II integral = $\int (y^2 - 4xy - 2x^2)dy = \frac{y^3}{3} + c_2$

[treating x constant and omitting the terms which are already present in I[st] integral]

$\therefore$ The required solution is: $\frac{x^3}{3} - 2x^2y - 2xy^2 + c_1 + \frac{y^3}{3} + c_2 = 0$

i.e., $x^3 + y^3 - 6xy(x+y) + c = 0$; where $c = 3(c_1 + c_2)$.

Example 3:

Solve the following differential equations:

i. $\frac{dy}{dx} + 3x^2y = 0$ *ii.* $\frac{dy}{dx} + y\tan x = \cos x$

iii. $\frac{dy}{dx} + 6y = 4$ *iv.* $\frac{dy}{dx} + 2xy = 3x$

Solution:

i. Given that $\frac{dy}{dx} + 3x^2y = 0$

It is in the linear form $\frac{dy}{dx} + Py = Q$; where $P = 3x^2$, and $Q = 0$.

$$\int P dx = \int 3x^2\, dx = x^3$$

$$\text{I.F} = e^{\int P dx} = e^{x^3}$$

Solution is : $ye^{\int P dx} = \int \left(e^{\int P dx}\right) Q dx + c$

i.e., $ye^{x^3} = 0 + c$

i.e., $\quad ye^{x^3} = c.$

ii. Given that $\dfrac{dy}{dx} + y\tan x = \cos x$

It is in the linear form $\dfrac{dy}{dx} + Py = Q$; where $P = \tan x$, and $Q = \cos x$.

$$\int P dx = \int \tan x \, dx = \log\sec x$$

$$\text{I.F} = e^{\int Pdx} = e^{\log\sec x} = \sec x$$

Solution is : $ye^{\int Pdx} = \int\left(e^{\int Pdx}\right)Qdx + c$

i.e., $\quad y\sec x = \int \sec x.\cos x \, dx + c$

i.e., $\quad y\sec x = \int dx + c$

iii. Given that $\dfrac{dy}{dx} + 6y = 4$

It is in the linear form $\dfrac{dy}{dx} + Py = Q$; where $P = 6$, and $Q = 4$.

$$\int P dx = \int 6 \, dx = 6x$$

$$\text{I.F} = e^{\int Pdx} = e^{6x}$$

Solution is : $ye^{\int Pdx} = \int\left(e^{\int Pdx}\right)Qdx + c$

i.e., $\quad ye^{6x} = \int e^{6x}.4 \, dx + c_1$

i.e., $\quad ye^{6x} = 4\dfrac{e^{6x}}{6} + c_1$

i.e., $e^{6x}(3y-2) = c$; where $3c_1 = c$

iv. Given that $\dfrac{dy}{dx} + 2xy = 3x$

It is in the linear form $\frac{dy}{dx}+Py=Q$; where $P=2x$, and $Q=3x$.

$$\int Pdx = \int 2x\,dx = x^2$$

$$\text{I.F} = e^{\int Pdx} = e^{x^2}$$

Solution is: $ye^{\int Pdx} = \int\left(e^{\int Pdx}\right)Qdx + c$

i.e., $ye^{x^2} = \int e^{x^2}.3x\,dx + c_1$

i.e., $ye^{x^2} = \frac{3}{2}e^{x^2} + c_1$

i.e., $e^{x^2}(2y-3)=c$; where $2c_1 = c$.

Example 4:

Solve the following differential equations:

i. $\frac{d^2y}{dx^2}-4\frac{dy}{dx}-5y=0$ *ii.* $\frac{d^2y}{dx^2}-6\frac{dy}{dx}+8y=0$

iii. $(D^2-6D+9)y=0$ *iv.* $\frac{d^2y}{dx^2}+2\frac{dy}{dx}+5y=0$

Solution:

i. The auxiliary equation is $m^2-4m-5=0$

Roots are m_1 = 5 and m_2 = –1

Complimentary Function, $CF = Ae^{5x}+Be^{-x}$

$\therefore$ General Solution is: $y=Ae^{5x}+Be^{-x}$

ii. The auxiliary equation is $m^2-6m+8=0$

Roots are m_1 = 4 and m_2 = 2

Complimentary Function, $CF = Ae^{4x}+Be^{2x}$

$\therefore$ General Solution is: $Ae^{4x}+Be^{2x}$

iii. The auxiliary equation is $m^2-6m+9=0$

Roots are m = 3, 3

Complimentary Function, $CF = (A+Bx)e^{3x}$

$\therefore$ General Solution is: $(A+Bx)e^{3x}$

iv. The auxiliary equation is $m^2+2m+5=0$

Roots are $m_1, m_2 = \alpha \pm i\beta = -1 \pm 2i$

Complimentary Function, $CF = e^{-x}(A\cos 2x + B\sin 2x)$

$\therefore$ General Solution is: $e^{-x}(A\cos 2x + B\sin 2x)$

Example 5(a):

Solve the following differential equations:

i. $(D^2+4D+3)y = e^{3x}$

ii. $(D^2-24D+144)y = e^{5x}+e^{-7x}$

iii. $\frac{d^2y}{dx^2}+5\frac{dy}{dx}+7y = e^{-x/2}+5$

iv. $\frac{d^2y}{dx^2}-3\frac{dy}{dx}+2y = e^{x}$

v. $(D^2-12D+36)y = e^{6x}$

Solution:

i. The auxiliary equation is $m^2+4m+3=0$

Roots are $m_1 = -3$ and $m_2 = -1$

Complimentary Function, $CF = Ae^{-3x}+Be^{-x}$

Particular Integral, $PI = \frac{1}{D^2+4D+3}e^{3x} = \frac{1}{(3)^2+4.3+3}e^{3x} = \frac{1}{24}e^{3x}$

$\therefore$ General Solution is: $y = Ae^{-x}+Be^{-3x}+\frac{1}{24}e^{3x}$

ii. The auxiliary equation is $m^2-24m+144=0$

$$m = 12,\ 12$$

$$\therefore \quad CF = (A+Bx)e^{12x}$$

$$PI_1 = \frac{1}{D^2-24D+144}e^{5x}$$

$$= \frac{1}{(5)^2-24.5+144}e^{5x} = \frac{1}{49}e^{5x}$$

$$PI_2 = \frac{1}{D^2-24D+144}e^{-7x}$$

$$= \frac{1}{(-7)^2 + 24.7 + 144} e^{-7x} = \frac{1}{361} e^{-7x}$$

$\therefore$ General Solution is:

$$y = CF + PI_1 + PI_2 = (A + Bx)e^{12x} + \frac{1}{49} e^{5x} + \frac{1}{361} e^{-7x}$$

iii. The auxiliary equation is $m^2 + 5m + 7 = 0$

$$m_1 , m_2 = \alpha \pm i\beta = \frac{-5}{2} \pm i \frac{\sqrt{3}}{2}$$

$$\therefore \quad CF = e^{-5x/2} \left(A \cos \frac{\sqrt{3}}{2} x + B \sin \frac{\sqrt{3}}{2} x \right)$$

$$PI_1 = \frac{1}{D^2 + 5D + 7} e^{-x/2}$$

$$= \frac{1}{(-1/2)^2 + 5(-1/2) + 7} e^{-x/2} = \frac{1}{17} e^{-x/2}$$

$$PI_2 = \frac{1}{D^2 + 5D + 7} (5)$$

$$= \frac{1}{D^2 + 5D + 7} (5e^{0x}) = \frac{5}{(0)^2 + 5(0) + 7} e^{0x} = \frac{5}{7}$$

$\therefore$ General Solution is: $y = CF + PI_1 + PI_2$

i.e., $$y = e^{-5x/2} \left(A \cos \frac{\sqrt{3}}{2} x + B \sin \frac{\sqrt{3}}{2} x \right) + \frac{1}{17} e^{-x/2} + \frac{5}{7}$$

iv. The auxiliary equation is $m^2 - 3m + 2 = 0$

Roots are $m_1 = 2$ and $m_2 = 1$

Complimentary Function, $CF = Ae^{2x} + Be^x$

Particular Integral, $PI = \frac{1}{D^2 - 3D + 2} e^x$

When we put D = 1, the denominator of PI becomes zero. (Hence the method fails).

Now we proceed as follows:

$$PI = \frac{1}{D^2 - 3D + 2} e^x = \frac{1}{(D-2)(D-1)} e^x$$

$$= \frac{1}{(D-1)} \frac{1}{(D-2)} e^x = \frac{1}{(D-1)} \frac{1}{(1-2)} e^x$$

$$= \frac{1}{(D-1)} \frac{1}{(1-2)} e^x = (-1)\frac{1}{(D-1)} e^x = -xe^x$$

$\therefore$ General Solution is: $y = Ae^{2x} + Be^x - xe^x$

v. The auxiliary equation is $m^2 - 12m + 36 = 0$

Roots are m = 6, 6

Complimentary Function, $CF = (A + Bx)e^{6x}$

Particular Integral, $PI = \frac{1}{D^2 - 12D + 36} e^{6x} = \frac{1}{(D-6)^2} e^{6x}$

When we put D = 6, the denominator of PI becomes zero. (Hence the method fails). Since 6 is a repeated root of the auxiliary equation, we have:

$$PI = \frac{1}{(D-6)^2} e^{6x} = \frac{x^2}{2!} e^{6x} = \frac{x^2 e^{6x}}{2}$$

$\therefore$ General Solution is: $y = (A + Bx)e^{6x} + \frac{x^2 e^{6x}}{2}$

Example 5(b):

Solve the following differential equations:

i. $(D^2 + 4)y = \cos 3x$

ii. $(D^2 + D + 1)y = \sin 2x$

iii. $\frac{d^2y}{dx^2} - 2\frac{dy}{dx} + y = e^{2x} - \cos x$

iv. $\frac{d^2y}{dx^2} + 6\frac{dy}{dx} + 9y = 2e^{-3x} + \sin 2x$

v. $(D^2 + 25)y = 2\sin 5x$

Solution:

i. The auxiliary equation is $m^2+4=0$

Roots are $m_1, m_2 = \pm 2i$

$$CF = A\cos 2x + B\sin 2x$$

$$PI = \frac{1}{D^2+4}\cos 3x = \frac{1}{-(3)^2+4}\cos 3x = \frac{-1}{5}\cos 3x$$

$\therefore$ General Solution is: $y = A\cos 2x + B\sin 2x + \frac{1}{5}\cos 3x$

ii. The auxiliary equation is $m^2+m+1=0$

Roots are $m_1, m_2 = \frac{-1\pm i\sqrt{3}}{2}$

$$CF = e^{-x/2}\left(A\cos\tfrac{\sqrt{3}}{2}x + B\sin\tfrac{\sqrt{3}}{2}x\right)$$

$$PI = \frac{1}{D^2+D+1}\sin 2x = \frac{1}{-(2)^2+D+1}\sin 2x$$

$$= \frac{1}{D-3}\sin 2x = \frac{D+3}{D^2-9}\sin 2x$$

$$= \frac{D+3}{-(2)^2-9}\sin 2x = \frac{D+3}{-13}\sin 2x$$

$$= \frac{-1}{13}[D(\sin 2x) + 3\sin 2x]$$

$$= \frac{-1}{13}[2\cos 2x + 3\sin 2x]$$

$\therefore$ General Solution is:

$$y = e^{-x/2}\left(A\cos\tfrac{\sqrt{3}}{2}x + B\sin\tfrac{\sqrt{3}}{2}x\right) + \frac{1}{13}[2\cos 2x + 3\sin 2x]$$

iii. The auxiliary equation is $m^2-2m+1=0$

Roots are $m = 1, 1$

$$CF = (A+Bx)e^x$$

$$PI_1 = \frac{1}{D^2-2D+1}e^{2x} = \frac{1}{2^2-2.2+1}e^{2x} = e^{2x}$$

$$PI_2 = \frac{1}{D^2-2D+1}\cos x$$

$$= \frac{1}{-(1)^2-2D+1}\cos x = \frac{-1}{2D}\cos x$$

$$= \frac{-1}{2}\int \cos x\, dx = \frac{-\sin x}{2}$$

$\therefore$ General Solution is: $y = (A+Bx)e^x + e^{2x} + \frac{\sin x}{2}$

iv. The auxiliary equation is $m^2+6m+9=0$

Roots are $m = -3, -3$

$$CF = (A+Bx)e^{-3x}$$

$$PI_1 = \frac{1}{D^2+6D+9}(2e^{-3x}) \text{ (method fails)}$$

$$= 2\frac{x^2}{2!}e^{-3x} = x^2e^{-3x}$$

$$PI_2 = \frac{1}{D^2+6D+9}(\sin 2x) = \frac{1}{-(2)^2+6D+9}(\sin 2x)$$

$$= \frac{1}{6D+5}(\sin 2x) = \frac{6D-5}{36D^2-25}(\sin 2x)$$

$$= \frac{-1}{169}(6D\sin 2x - 5\sin 2x)$$

$$= \frac{-1}{169}(12\cos 2x - 5\sin 2x)$$

$\therefore$ General Solution is: $y = CF + PI_1 + PI_2$

$$= (A+Bx)e^{-3x} + x^2e^{-3x} - \frac{1}{169}(12\cos 2x - 5\sin 2x)$$

v. The auxiliary equation is $m^2+6m+9=0$

Roots are $m = \pm 5i$

$$CF = A\cos 5x + B\sin 5x$$

$$PI = \frac{1}{D^2+25}(\sin 2x) \text{ (method fails)}$$

$$= \frac{1}{D^2+25}(\text{I.P of } e^{i5x})$$

$$= \text{I.P of } \frac{1}{(D-i5)(D+i5)}(e^{i5x})$$

$$= \text{I.P of } \frac{-i}{10}\frac{1}{(D-i5)}(e^{i5x})$$

$$= \text{I.P of } \frac{-i}{10}x(e^{i5x})$$

$$= \text{I.P of } \frac{-i}{10}x(\cos 5x + i\sin 5x)$$

$$= \text{I.P of } \frac{-i}{10}x(\cos 5x + i\sin 5x)$$

$\therefore$ General Solution is:

$$y = CF + PI_1 + PI_2$$

$$= (A+Bx)e^{-3x} + x^2e^{-3x} - \frac{1}{169}(12\cos 2x - 5\sin 2x)$$

Example 5(c):

Solve the following differential equations:

i. $(D^2+2D+1)y = 2x + x^2$

ii. $(D^2-D-2)y = 44-76x$

iii. $6\frac{d^2y}{dx^2} - \frac{dy}{dx} - 2y = e^{4x} + x^2$

iv. $(D^2+2D+1) = 2e^{3x} + x^2$

v. $(D^2+3D+2)y = e^{-x} + x^2 + \sin x$

Solution:

i. The auxiliary equation is $m^2+2m+1=0$

Roots are $m = -1, -1$

$$CF = (A+Bx)e^{-x}$$

$$PI = \frac{1}{D^2+2D+1}(2x+x^2)$$

$$= \frac{1}{(1+D)^2}(2x+x^2) = (1+D)^{-2}(2x+x^2)$$

$$= (1-2D+3D^2-....)(2x+x^2)$$

$$= (2x+x^2)-2D(2x+x^2)+3D^2(2x+x^2);;$$

dropping powers of D higher than 2

$$= (2x+x^2)-2(2+2x)+3(2)$$

$$= x^2-2x+2$$

$\therefore$ General Solution is: $y = (A+Bx)e^{-x}+x^2-2x+2$

ii. The auxiliary equation is $m^2-m-2=0$

Roots are $m_1 = 2,\ m_2 = -1$

$$CF = Ae^{2x}+Be^{-x}$$

$$PI = \frac{1}{D^2-D-2}(44-76x)$$

$$= \frac{1}{-2(1+\frac{1}{2}D-\frac{1}{2}D^2)}(44-76x)$$

$$= \frac{-1}{2}\left[1+\left(\frac{D-D^2}{2}\right)\right]^{-1}(44-76x)$$

$$= \frac{-1}{2}\left[1-\left(\frac{D-D^2}{2}\right)+.....\right](44-76x);$$

dropping higher powers of D

$$= \frac{-1}{2}\left[44-76x-\frac{1}{2}(-76)\right]$$

$$= \frac{-1}{2}[82-76x] = 38x-41$$

$\therefore$ General Solution is: $y = Ae^{2x} + Be^{-x} + 38x - 41$

iii. The auxiliary equation is $6m^2 - m - 2 = 0$

Roots are $m_1 = 2/3$, $m_2 = -1/2$

$$CF = Ae^{\frac{2x}{3}} + Be^{\frac{-x}{2}}$$

$$PI_1 = \frac{1}{6D^2 - D - 2}(e^{4x}) = \frac{e^{4x}}{90}$$

$$PI_2 = \frac{1}{6D^2 - D - 2}(x^2)$$

$$= \frac{1}{-2(1 + \frac{1}{2}D - 3D^2)}(x^2)$$

$$= \frac{-1}{2}\left[1 + \left(\frac{D}{2} - 3D^2\right)\right]^{-1}(x^2)$$

$$= \frac{-1}{2}\left[1 - \left(\frac{D}{2} - 3D^2\right) + \left(\frac{D}{2} - 3D^2\right)^2 - \ldots\ldots\right](x^2)$$

$$= \frac{-1}{2}\left[x^2 - \left(\frac{D}{2} - 3D^2\right)(x^2) + \left(\frac{D^2}{4}\right)(x^2)\right]$$

dropping higher powers of D

$$= \frac{-1}{2}\left[x^2 - \tfrac{1}{2}.2x + 3.2 + \tfrac{1}{4}.2\right] = \frac{-1}{4}\left[2x^2 - 2x + 13\right]$$

$\therefore$ General Solution is: $y = Ae^{\frac{2x}{3}} + Be^{\frac{-x}{2}} + \dfrac{e^{4x}}{90} - \dfrac{1}{4}(2x^2 - 2x + 13)$

iv. The auxiliary equation is $m^2 + 2m + 1 = 0$

Roots are $m = -1, -1$

$$CF = (A + Bx)e^{-x}$$

$$PI_1 = \frac{1}{D^2 + 2D + 1}(2e^{3x}) = \frac{e^{3x}}{8}$$

$$PI_2 = \frac{1}{D^2 + 2D + 1}(x^2)$$

$$= \frac{1}{(1+D)^2}(x^2) = [1+D]^{-2}(x^2)$$

$$= \left[1-2D+3D^2-......\right](x^2) = \left[x^2-2.2x+3.2\right];$$

dropping higher powers of D

$$= x^2-4x+6$$

$\therefore$ General Solution is: $y=(A+Bx)e^{-x}+\dfrac{e^{3x}}{8}+x^2-4x+6$

v. The auxiliary equation is $m^2+3m+2=0$

Roots are m_1 , m_2 = –1, –2

$$CF = Ae^{-2x}+Be^{-x}$$

$$PI_1 = \frac{1}{D^2+3D+2}(e^{-x}) = xe^{-x}$$

$$PI_2 = \frac{1}{D^2+3D+2}(x^2)$$

$$= \frac{1}{2(1+\frac{3}{2}D+\frac{1}{2}D^2)}(x^2) = \frac{1}{2}\left[1+\frac{3D+D^2}{2}\right]^{-1}(x^2)$$

$$= \frac{1}{2}\left[1-\frac{3D+D^2}{2}+\left(\frac{3D+D^2}{2}\right)^2-.....\right](x^2)$$

$$= \frac{1}{2}\left[x^2-\tfrac{1}{2}(3.2x+2)+\tfrac{1}{4}(9.2)\right] = \tfrac{1}{4}(2x^2-6x+7)$$

$$PI_3 = \frac{1}{D^2+3D+2}(\sin x)$$

$$= \frac{1}{-(1)^2+3D+2}(\sin x) = \frac{1}{3D+1}(\sin x)$$

$$= \frac{3D-1}{9D^2-1}(\sin x) = \frac{3\cos x-\sin x}{-10}$$

$$= \frac{\sin x-3\cos x}{10}$$

$\therefore$ General Solution is:

$$y = Ae^{-2x} + Be^{-x} + xe^{-x} + \tfrac{1}{4}(2x^2 - 6x + 7) + \frac{\sin x - 3\cos x}{10}$$

Example 6:

Find the demand function, if the elasticity of demand is given by: a/bx-1 where a and b are positive constants.

Solution:

Given that $e = \dfrac{a}{bx} - 1$

i.e., $-\dfrac{p}{x}\dfrac{dx}{dp} = \dfrac{a}{bx} - 1$

i.e., $-\dfrac{p}{x}\dfrac{dx}{dp} = \dfrac{a}{bx} - 1$

This is a differential equation in the variable separable form. On separating the variables we get,

$$\frac{dp}{p} = \frac{bdx}{bx - a}$$

Its solution is:

$$\int\frac{dp}{p} = \int\frac{bdx}{bx - a}$$

i.e., $\log p = \log(bx - a) + c_1$

i.e., $\log p = \log c(bx - a)$; where $c_1 = \log c$

i.e., $p = c(bx - a)$

i.e., $x = \dfrac{p + ac}{bc}$

Example 7:

The relationship between net profit P and advertising expenditure x is such that the rate of increase of net profit as advertising expenditure increases is proportional to a constant minus net profit. Find the relationship between net profit and advertising expenditure if $P = P_0$ when $x = 0$.

Solution:

We have, $\dfrac{dP}{dx} = k(a - P)$;

where a and k are constants Separating the variables we get, $\frac{dP}{a-P} = kdx$

Its solution is: $\int \frac{dP}{a-P} = k\int dx$

i.e., $-\log(a-P) = kx + c_1$

i.e., $a - P = ce^{-kx}$

i.e., $P = a - ce^{-kx}$

Substituting the initial conditions: $P_0 = a - c$ or $c = a - P_0$

$\therefore$ The required relationship is: $P = a - (a - P_0)e^{-kx}$

Example 8(a):

The rate of increase of sales volume s as price p decreases is proportional to sales volume and inversely proportional to price minus a constant. Find the relationship between sales volume and price, if $s = s_0$ when $p = p_0$.

Solution:

We have, $\frac{ds}{dp} = -\frac{as}{p-b}$; where a and b are constantsseparating the variables we get, $\frac{ds}{s} = -\frac{a}{p-b}dp$

Its solution is: $\int \frac{ds}{s} = -a\int \frac{1}{p-b}dp$

i.e., $-\log s = -a\log(p-b) + c_1$

i.e., $s(p-b)^a = c$

i.e., $s = \frac{c}{(p-b)^a}$

Using initial conditions, $s_0 = \frac{c}{(p_0 - b)^a}$, or $c = s_0(p_0 - b)^a$

$\therefore$ The required relationship is: $s = s_0\left(\frac{p_0 - b}{p-b}\right)^a$

Example 8(b):

The change in price y with change in quantity demanded x of a particular commodity is given by, $\frac{dy}{dx} = -\frac{2xy+24x}{x^2+16}$*. Find the relationship between price and quantity demanded if the price is 7.5 when the quantity demanded is 4.*

Solution:

We have, $\frac{dy}{dx} = -\frac{2xy+24x}{x^2+16}$

i.e., $(2xy+24x)dx+(x^2+16)dy=0$.

This is in the form $P(x,y)dx+Q(x,y)dy=0$; where

$$P = 2xy+24x, \; Q = x^2+16$$

$$\frac{\partial P}{\partial y} = 2x = \frac{\partial Q}{\partial x}$$

Hence the given differential equation is exact.

I integral $= \int(2xy+24x)dx = 2\frac{x^2}{2}y+24\frac{x^2}{2}+c_1$ [treating y constant]

$= x^2y+12x^2+c_1$ [treating y constant]

II integral $= \int(x^2+16)dy = 16y+c_2$

[treating x constant and omitting the terms which are already present in I[st] integral]

∴ The required solution is:

$$x^2y+12x^2+c_1+16y+c_2=0$$

i.e., $12x^2+x^2y+16y=c$

Using initial conditions, we get $c = 432$

∴ The required relationship is: $12x^2+x^2y+16y=432$

Example 8(c):

In Domar's model, the national income Y and public debt D are connected by the equations:

$$\frac{dY}{dt} = \beta Y \text{ and } \frac{dD}{dt} = \alpha Y \quad (\alpha, \beta \text{ constants})$$

Prove that the ratio of debt to income approaches constant.

Solution:

$$\frac{dY}{dt} = \beta Y \Rightarrow \frac{dY}{Y} = \beta dt$$

$$\Rightarrow \quad \int \frac{dY}{Y} = \int \beta dt$$

$$\Rightarrow \quad \log Y = \beta t + k_1; \; k_1 \text{ constant}$$

$$\Rightarrow \quad Y = e^{\beta t + k_1}$$

$$\Rightarrow \quad Y = c_1 e^{\beta t} \text{ ; where } c_1 = e^{k_1}$$

$$\Rightarrow \quad \frac{dD}{dt} = \alpha Y \Rightarrow dD = \alpha Y dt$$

$$\Rightarrow \quad \frac{dD}{dt} = \alpha Y \Rightarrow dD = \alpha c_1 e^{\beta t} dt$$

$$\Rightarrow \int dD = \alpha c_1 \int e^{\beta t} dt$$

$$\Rightarrow \quad D = \alpha c_1 \frac{e^{\beta t}}{\beta}$$

$$\Rightarrow \quad D = c_2 e^{\beta t} \text{ ; where } c_2 = \frac{c_1 \alpha}{\beta}$$

Hence, $\frac{D}{Y} = \frac{c_2 e^{\beta t}}{c_1 e^{\beta t}} = \frac{c_2}{c_1}$, a constant

Example 9:

The price elasticity of demand curve $x = f(p)$ *is of the form* $a - bp$, *where a, b constants. Find the demand curve.*

Solution:

Given that, elasticity of demand, $e = a - bp$

i.e., $\frac{-x}{p} \frac{dp}{dx} = a - bp$

i.e., $-xdp = (a-bp)pdx$

i.e $\dfrac{-dp}{(a-bp)p} = \dfrac{dx}{x}$

Now, by the method of partial fractions,

$$\frac{-1}{(a-bp)p} = \frac{-b/a}{a-bp} + \frac{-1/a}{p}$$

$$\therefore \quad \frac{-dp}{(a-bp)p} = \frac{dx}{x} \Rightarrow \left[\frac{-b/a}{a-bp} + \frac{-1/a}{p}\right]dp = \frac{dx}{x}$$

$$\Rightarrow \quad \int\left[\frac{-b/a}{a-bp} + \frac{-1/a}{p}\right]dp = \int\frac{dx}{x}$$

$$\Rightarrow \quad \frac{-b}{a}\log(a-bp).\frac{-1}{b} - \frac{1}{a}\log p = \log x + \log c$$

$$\Rightarrow \quad \frac{-1}{a}\left[\log(a-bp) - \log p\right] = \log xc$$

$$\Rightarrow \quad \left[\log p - \log(a-bp)\right] = a\log xc$$

$$\Rightarrow \quad \frac{p}{a-bp} = (xc)^a$$

$$\Rightarrow \quad \frac{p}{a-bp} = (xc)^a$$

$$\Rightarrow \quad p\left[(xc)^a + b\right] = a$$

$$\Rightarrow \quad p = \frac{a}{(xc)^a + b}, \text{ which is the demand curve.}$$

Example 10:

The change in the net profit P, as advertising expenditure (x) changes, is given by $\dfrac{dP}{dx} = 3 - 2(P+x)$. *Find P if P = 10 at x = 10.*

Solution:

Given $\dfrac{dP}{dx} = 3 - 2(P+x) = 3 - 2P - 2x$

$$\Rightarrow \qquad \frac{dP}{dx}+2P=3-2x$$

This is a linear differential equation in the form $\frac{dy}{dx}+Py=Q.$

Hence the solution is: $Pe^{\int 2dx} = \int (3-2x)e^{\int 2dx}dx$

$$\Rightarrow \qquad Pe^{2x} = \int (3-2x)e^{2x}dx$$

$$= (3-2x)\frac{e^{2x}}{2}+\int -2\frac{e^{2x}}{2}dx$$

$$= (3-2x)\frac{e^{2x}}{2}+\frac{e^{2x}}{2}+c$$

$$\Rightarrow \qquad P = 2-x+ce^{2x}$$

Given, $\qquad P = 10$ at $x = 10$

i.e., $\qquad 10 = -8+ce^{20}$

$$\Rightarrow \qquad c = 18e^{-20}$$

$$\therefore \qquad P = 2-x+18e^{2x-20}$$

Example 11(a):

Show that the Wronskion of the function x^2, $x^2 \log x$ is non-zero can these function be independent solutions of an ordinary differential equations, if so determine this equation.

Solution:

Let $y_1 = x^2$ and $y_2 = x^2 \log x$

Wronskian of these function is

$$W(x) = \begin{vmatrix} y_1 & y_2 \\ y_1' & y_2'' \end{vmatrix}$$

$$= \begin{vmatrix} x^2 & x^2 \log x \\ 2x & 2x \log x \end{vmatrix} = x^2 \neq 0$$

$\therefore$ These functions y_1 and y_2 are linearly independent solutions of an ordinary differential equation.

To Determine the Equation:

Let $y = c_1y_1 + c_2y_2 = c_1x^2 + c_2x^2 \log x$

so that $y_1 = 2c_1 + 2c_2 \log x + 3\ c_2$

Eliminating c_1, c_2 from the above three relations, we get

$$x^2y_2 - 2xy_1 + 4y = 0$$

which is the required differential equation. **Ans.**

Example 11(b):

Show that $y_1 = \sin x$ and $y_2(x) = \sin(x) = \sin x - \cos x$ are linearly independent solutions $y'' + y = 0$.

Solution:

Let $y_1 = \sin x$ and $y_2(x) = \sin(x) = \sin x - \cos x$ satisfy the differential equation $y'' + y = 0$.

∴ They are solutions of the equation

Now, Wronskian of these solutions y_1 and y_2 is

$$W(x) = \begin{vmatrix} y_1 & y_2 \\ y_1' & y_2'' \end{vmatrix}$$

$$= \begin{vmatrix} \sin x & \sin x - \cos x \\ \cos x & \cos x + \sin x \end{vmatrix}$$

$$= \sin x\ (\cos x + \sin x) - \cos x\ (\sin x - \cos x)$$

$$= 1 \neq 0.$$

∴ These $y_1(x)$ and $y_2(x)$ are linearly independent solutions of $y'' + y = 0$.

Example 12:

Find the third order differential equation whose solution is the 3-parameter family of curves defined by

$$x^2 + y^2 + 2ax + 2by + c = 0 \qquad ...(1)$$

Solution:

Differentiating three times is succession

$$x + yy^{(1)} + a + by^{(1)} = 0 \qquad ...(2)$$

$$1 + (y^{(1)})^2 + yy^{(2)} + by^{(2)} = 0 \qquad ...(3)$$

$$3y^{(1)}\ y^{(2)} + yy^{(3)} + by^{(3)} = 0 \qquad ...(4)$$

Eliminating b from equations (4) and (3)

$$\{1 + (y^{(1)})^2\} y^{(3)} - 3y^{(1)} (y^{(3)})^2 = 0.$$

Example 13:

Eliminate the constant a from

$$\sqrt{1-x^2} + \sqrt{1-y^2} = a(x-y) \qquad ...(1)$$

Solution:

Put $x = \sin\theta$

$y = \sin\phi$

$\therefore$ Equation (1) reduces to

$$\sqrt{1-\sin^2\theta} + \sqrt{1-\sin^2\phi} = a(\sin\theta - \sin\phi)$$

$$\cos\theta + \cos\phi = a(\sin\theta - \sin\phi)$$

$$\Rightarrow \quad 2\cos\frac{\theta+\phi}{2}\cos\frac{\theta-\phi}{2} = 2a\cos\frac{\theta+\phi}{2}\sin\frac{\theta-\phi}{2}$$

$$\Rightarrow \quad \cot\frac{\theta-\phi}{2} = a$$

$$\Rightarrow \quad \theta = \phi = \cot^{-1}a$$

$$\Rightarrow \quad \sin^{-1}x - \sin^{-1}y = 2\cot^{-1}a$$

Differentiating once, we get

$$\frac{1}{\sqrt{1-x^2}} - \frac{1}{\sqrt{1-y^2}} y^{(1)} = 0.$$

Example 14(a):

Show that the Wronskian of e^{ax} cos bx and e^{ax} sin bx (b ' 0) is be^{2ax}.

Solution:

Let $y_1 = e^{ax}\cos bx$ and $y_2 = e^{ax}\sin bx$.

Wronskian of y_1, y_2 is

$$W(x) = \begin{vmatrix} y_1 & y_2 \\ y'_1 & y'_2 \end{vmatrix}$$

$$= \begin{vmatrix} e^{ax}\cos bx & e^{ax}\sin bx \\ e^{ax}(a\cos bx - b\sin bx) & e^{ax}(a\sin bx + b\cos bx) \end{vmatrix} = be^{2ax}$$

Example 14(b):

Prove that the Wronskian of the functions.

$$e^m\, e^{m_1x}, e^{m_2x}, e^{m_3x} \text{ is equal}$$

to $(m_1 - m_2)(m_2 - m_3)(m_3 - m_1).\, e^{(m_1+m_2+m_3)x}$.

Solution:

Let $y_1 = e^{m_1x}, y_2 = e^{m_2x}, y_3 = e^{m_3x}$,

$\therefore$ Wronskian of y_1, y_2, y_3, is

$$W(x) = \begin{vmatrix} y_1 & y_2 & y_3 \\ y'_1 & y'_2 & y'_3 \\ y''_1 & y''_2 & y''_3 \end{vmatrix} = \begin{vmatrix} e^{m_1x} & e^{m_2x} & e^{m_3x} \\ m_1e^{m_1x} & m_2e^{m_2x} & m_3e^{m_3x} \\ m_1^2e^{m_1x} & m_2^2e^{m_2x} & m_3^2e^{m_3x} \end{vmatrix}$$

$$= \begin{vmatrix} 1 & 1 & 1 \\ m_1 & m_2 & m_3 \\ m_1^2 & m_2^2 & m_3^2 \end{vmatrix} e^{m_1x}.e^{m_2x}.e^{m_3x}$$

$$= (m_1 - m_2)(m_2 - m_3)(m_3 - m_1) \cdot e^{(m_1+m_2+m_3)x}$$ **Ans.**

Example 15:

Which of the following sets of functions are linearly independent.

(i) $\sin x, \cos x, \sin 2x$

(ii) $1 + x, 1 + 2x, x^2$

(ii) $x^2 - x + 1, x^2 - 1, 3x^2 - x - 1$.

Solution:

(i) Let $y_1 = \sin x, y_2 = \cos x, y_3 = \sin 2x$.

Wronskian of these functions is

$$W(x) = \begin{vmatrix} y_1 & y_2 & y_3 \\ y'_1 & y'_2 & y'_3 \\ y''_1 & y''_2 & y''_3 \end{vmatrix} = \begin{vmatrix} \sin x & \cos x & \sin 2x \\ \cos x & -\sin x & 2\cos 2x \\ -\sin x & -\cos x & -4\sin 2x \end{vmatrix}$$

$$= \sin x \begin{vmatrix} -\sin x & 2\cos 2x \\ -\cos x & -4\sin 2x \end{vmatrix} - \cos x \begin{vmatrix} \cos x & \sin 2x \\ -\cos x & -4\sin 2x \end{vmatrix}$$

$$- \sin x \begin{vmatrix} \cos x & \sin 2x \\ -\sin x & 2\cos 2x \end{vmatrix}$$

$= \sin x.(4 \sin x \sin 2x + 2 \cos x \cos 2x)$

$\quad - \cos x (-4 \cos x \sin 2x + \cos \sin 2x)$

$\quad - \sin x (2 \cos x \cos 2x + \sin x \sin 2x)$

$= 3 \sin^2 x \sin 2x + 3 \cos^2 x \sin 2x + 2 \sin x \cos x \cos 2x$

$\quad - 2 \sin x \cos x \cos 2x$

$= 3 (\sin^2 x + \cos^2 x) \sin 2x + \sin 2x \cos 2x - \sin 2x \cos 2x$

$= 3 \sin 2x \neq 0.$

(ii) Here $y_1 = 1 + x$, $y_2 = 1 + 2x$, $y_3 = x^2$

$\therefore$ Wronskian $W(x) = \begin{vmatrix} 1+x & 1+2x & x^2 \\ 1 & 2 & 2x \\ 0 & 0 & 2 \end{vmatrix}$

$= 2 [2 (1 + x) - (1 + 2x)] = 2 \neq 0.$

$\therefore$ The set of functions is linearly independent.

(iii) Here $y_1 = x^2 - x + 1$, $y_2 = x^2 - 1$, $y_3 = 3x^2 - x - 1$

$\therefore$ Wronskian $W(x) = \begin{vmatrix} x^2 - x + 1 & x^2 - 1 & 3x^2 - x - 1 \\ 2x - 1 & 2x & 6x - 1 \\ 2 & 2 & 6 \end{vmatrix}$

$= (x^2 - x + 1) \{12x - 2(6x - 1)\} - (x^2 - 1) \{6(2x - 1) - 2(6x - 1)\}$
$\quad + (3x^2 - x - 1) \{2 (2x - 1) - 4x\} = 2(x^2 - x + 1)$
$\quad + 4(x^2 - 1) - 2(3x^2 - x - 1) = 0$

$\therefore$ The set of functions is not linearly independent.

Example 16:

Use Wronskian to show that the functions, x, x^2, x^3 are independent. Determine the differential equation with these as independent solutions.

Solution:

Let $y_1 = x$, $y_2 = x^2$ and $y_3 = x^3$.

Wronskian of these function is

$$W(x) = \begin{vmatrix} y_1 & y_2 & y_3 \\ y'_1 & y'_2 & y'_3 \\ y''_1 & y''_2 & y''_3 \end{vmatrix} = \begin{vmatrix} x & x^2 & x^3 \\ 1 & 2x & 3x^2 \\ 0 & 2 & 6x \end{vmatrix}$$

$= x(12x^2 - 6x^2) - 1(6x^3 - 2x^3) = 2x^3 \neq 0$

$\therefore$ The given functions are independent.

To determine the equation.

Let $\quad y = c_1y_2 + c_2y_2 + c_3y_3 = c_1x + c_2x^2 + c_3x^3 \qquad ...(1)$

$$\frac{dy}{dx} = c_1 + 2c_2x + 3c_3x^2 \qquad ...(2)$$

$$\frac{d^2y}{dx^2} = 2c_2 + 6c_3x \qquad ...(3)$$

and $\quad \frac{d^3y}{dx^3} = 6c_3 \qquad ...(4)$

From (4), $c_3 = \frac{1}{6}\frac{d^3y}{dx^3}$.

From (3), $2c_2 = \frac{d^2y}{dx^2} - 6c_3x = \frac{d^2y}{dx^2} - x\frac{d^3y}{dx^3}$

$$\therefore c_2 = \frac{1}{2}\frac{d^2y}{dx^2} - \frac{1}{2}x\frac{d^3y}{dx^3}$$

Subtracting x times of (2), from (1), we get

$$y - x\frac{dy}{dx} = -c_2x^2 - 2c_3x^3$$

$$= -\left(\frac{1}{2}\frac{d^2y}{dx^2} - \frac{1}{2}x\frac{d^3y}{dx^3}\right)x^2 - 2\left(\frac{1}{6}\frac{d^3y}{dx^3}\right)x^3$$

$$x^3\frac{d^3y}{dx^3} - 3x^3\frac{d^2y}{dx^2} + 6x\frac{dy}{dx} - 6y = 0$$

Which is the required differential equations. **Ans.**

Example 17(a):

Show that $y = 1/4 \sin 4x$ is a unique solution of the initial value problem $y'' + 16y = 0$ with $y(0) = 0$, $y'(0) = 1$.

Solution:

We have $y' = \cos 4x$, $y'' = -4 \sin 4x$.

Now, $y'' + 16y = -4 \sin 4x + 16\left(\frac{1}{4}\sin 4x\right) = 0$.

Thus, $y = \frac{1}{4} \sin 4x$ is a solution of $y'' + 16y = 0$, where

$y(0) = 0,\ y'(0) = 1.$

Since $a_0(x) = 1 \neq 0$ for each x in $(-\infty, \infty)$, it follows from Fundamental Existence Theorem that, $y = 1/4 \sin 4x$ is a unique solution of $y'' + 16y = 0$ with $y(0) = 0,\ y'(0) = 1$.

Example 17(b):

Show that the Wronskian of the functions x^2, x^2 long x is non-zero. Can these functions be independent solutions of an ordinary differential equation; if so determine this equation.

Solution:

Let $y_1 = x^2$ and $y_2 = x^2 \log x$.

Wronskian of these functions is

$$W(x) \begin{vmatrix} y_1 & y_2 \\ y'_1 & y'_2 \end{vmatrix} = \begin{vmatrix} x^2 & x^2 \text{lox}\, x \\ 2x & 2x \log x + x \end{vmatrix} = x^3 \neq 0$$

∴ These functions y_1 and y_2 are linearly independent solutions of an ordinary differential equation.

To determine the equation

Let $y = c_1 y_1 + c_2 y_2 = c_1 x^2 + c_2 x^2 \log x$

so that $dy/dx = 2c_1 x + 2c_2 x \log x + c_2 x$

and $d^2y/dx^2 = 2c_1 + 2c_2 \log x + 3c_2$

Eliminating c_1, c_2 from the above three relations, we have

$$x^2\, d^2y/dx^2 - 3x\,(dy/dx) + 4y = 0$$

which is the required differential equation.

Example 18:

Show that the functions $y = cx^2 + x + 3$ is a solution, though not unique, of the initial value problem

$$x^2y'' - 2\,x\,y' + 2y = 6$$

with $y(0) = 3,\ y'(0) = 1$ *on* $(-\infty, \infty)$.

Solution:

We have $y = c\,x^2 + x + 3$, $y' = 2\,cx + 1$. $y'' = 2c$.

$$\therefore \quad x^2y'' - 2xy' + 2y = x^2(2c) - 2x(2cx + 1) + 2(cx^2 + x + 3)$$

$$= 4\,cx^2 - 4cx^2 - 2x + 2x + 6 = 6.$$

Thus, $y = cx^2 + x + 3$ is a solution of $x^2y'' - 2xy' + 2y = 6$, where $y(0) = c \times 0 + 0 + 3 = 3$, $y'(0) = 2c \times 0 + 1 = 1$.

Here $a_0(x) = x^2$, $a_1(x) = -2x$, $a_2(x) = 2$, $R(x) = 6$ are continuous functions on $(-\infty, \infty)$

It may be noted that $a_0(x) = 0$ for $x = 0 \in (-\infty, \infty)$.

Hence the solution $y = cx^2 + x + 3$ is not unique, as $y = cx^2 + x + 3$ is a solution for any real value of c.

Notice that $y = x^2 + x + 3$, $y = 2x^2 + x + 3$ are both solutions of $x^2y'' - 2xy' + 2y = 6$ with $y(0) = 3$, $y'(0) = 1$.

Example 19:

Find the differential equation of all tangent lines to the parabola

$$y = x^2.$$

Solution:

Parametric equations of the given parabola are

$$x = t, \qquad y = t^2$$

∴ The equation of any tangent is

$$2xt = y + t^2 \qquad \ldots(1)$$

Equation (1) is the family of tangent lines to the parabola, where t is the parameter.

Differentiating (1) once

$$2t = y^{(1)} \qquad \ldots(2)$$

Eliminating t between (1) and (2)

$$4(y - xy^{(1)}) + (y^{(1)})^2 = 0$$

Form the differential equations for the following families of curves.

Example 20(a):

Show that $y_1(x) = \sin x$ and $y_2(x) = \sin x - \cos x$ are linearly independent solutions of $y'' + y = 0$.

Solution:

Both $y_1(x) = \sin x$ and $y_2(x) = \sin x - \cos x$ satisfy the differential equation $y'' + y = 0$.

∴ They are solutions of the equation.

Now Wronskian of these solutions y_1 and y_2 is

$$W(x) = \begin{vmatrix} y_1 & y_2 \\ y'_1 & y'_2 \end{vmatrix} = \begin{vmatrix} \sin x & \sin x - \cos x \\ \cos x & \cos x + \sin x \end{vmatrix}$$

$= \sin x (\cos x + \sin x) - \cos x(\sin x - \cos x) = 1 \neq 0.$

$\therefore$ These $y_1(x)$ and $y_2(x)$ are linearly independent solutions of $y^n + y = 0$.

Theorem 1:

There exist two linearly independent solutions y_1 (x) and y_2 (x) of the equation

$$a_0 (x) y'' + a_1 (x) y' + a_2 (x) y = 0 \quad ...(1)$$

such that its every solution y(x) may be written as

$$y (x) = c_1 y_1 (x) + c_2 y_2 (x), x \in (a, b)$$

where c_1 and c_2 are suitably chosen constants.

Proof:

Let $x_0 \in (a, b)$. Let y_1 (x) and y_2 (x) be two solutions of (1) satisfying

$$y_1 (x_0) = 1 \text{ and } y_1' (x_0) = 0;$$

$$y_2 (x_0) = 0 \text{ and } y_2' (x_0) = 1. \quad ...(2)$$

First of all, we show that y_1 (x) and y_2 (x) are linearly independent. Let, if possible, they be linearly dependent. Then there exist constants c_1 and c_2, not both zero, such that

$$c_1y_1 (x) + c_2y_2 (x) \equiv 0 \quad \text{on} \quad (a, b)$$

$$\Rightarrow \quad c_1y_1' (x) + c_2y_2' (x) \equiv 0 \quad \text{on} \quad (a, b).$$

Since $x_0 \in (a, b)$, therefore

$$c_1y_1 (x_0) + c_2 y_2 (x_0) = 0$$

and $$c_1y_1' (x_0) + c_2 y_2' (x_0) = 0 \quad ...(3)$$

Using (2) in (3), we get $c_1 = 0$ and $c_2 = 0$. This is a contradiction to the fact that c_1 and c_2 are not both zero. Now we prove the second part of the theorem. Let y(x) be an arbitrary solution of (1) satisfying

$$y(x_0) = c_1 \text{ and } y' (x_0) = c_2. \quad ...(4)$$

Let $$Y(x) = y (x) - c_1 y_1 (x) - c_2 y_2 (x). \quad ...(5)$$

They Y (x) being a linear combination of solutions y(x), y_1 (x) and y_2 (x) is a solution of (1).

Now $Y (x_0) = y (x_0) - c_1 y_1 (x_0) - c_2 y_2 (x_0) = 0$, using (2) and (4)

and $Y' (x_0) = y' (x_0) - c_1 y_1' (x_0) - c_2y_2' (x_0) = 0$, using (2) and (4).

Thus, Y(x) is a solution of (1) such that $Y (x_0) = 0, Y' (x_0) = 0$ ...(6)

By Corollarly 1, $Y (x) \equiv 0$ on (a, b) and so by (5),

$$y (x) = c_1 y_1 (x) + c_2 y_2 (x)$$

where c_1 and c_2 are given by (4), This proves the theorem.

Theorem 2:

If $y_1(x)$ and $y_2(x)$ are any two solutions of

$$a_0 (x) y'' + a_1 (x) y' + a_2 (x) y = 0$$

then the linear combination $c_1 y_1 (x) + c_2 y_2 (x)$, where c_1 and c_2 are constants, is also a solution of the given equation.

Proof:

We have

$$a_0 (x) y_1'' (x) + a_1 (x) y_1' (x) + a_2 (x) y_1 (x) = 0 \quad \text{...(1)}$$

and $$a_0 (x) y_2'' (x) + a_1 (x) y_2' (x) + a_2(x) y_2 (x) = 0 \quad \text{...(2)}$$

Let $y (x) = c_1 y_1 (x) + c_2 y_2 (x)$.

Then $y'(x) = c_1 y_1' (x) + c_2 y_2' (x)$

and $y''(x) = c_1 y_1'' (x) + c_2 y_2'' (x)$

Consider $a_0 (x) y'' (x) + a_1 (x) y' (x) + a_2 (x) y (x)$

$$= c_1 [a_0 (x) y_1'' (x) + a_1 (x) y_1' (x) + a_2 (x) y_1 (x)]$$
$$+ c_2 [a_0 (x) y_2'' (x) + a_1 (x) y_2' (x) + a_2 (x) y_2 (x)].$$

using (3)

$= c_1, 0 + c_2 . 0$, using (1) and (2).

Hence, $y (x) = c_1 y_1 (x) + c_2 y_2 (x)$ is also a solution of the given equation.

Example 20(b):

Show that linearly independent solutions of

$$y'' - 3y' + 2y = 0 \text{ are } e^x \text{ and } e^{2x}.$$

Find the solution y (x) with the property that y (0) = 0, y' (0) = 1.

Solution:

The given equation is $(D^2 - 3D + 2) y = 0$, $(D = d/dx)$

$\Rightarrow$ $$(D - 1) (D - 2) y = 0.$$

Its solutions are e^x and e^{2x}.

Let $y_1 = e^x$ and $y_2 = e^{2x}$. Their Wronskian is

$$w(x) = \begin{vmatrix} y_1 & y_2 \\ y_1' & y_2' \end{vmatrix} = \begin{vmatrix} e^x & 2e^x \\ e^x & 2e^{2x} \end{vmatrix}$$

$$= 2e^x e^{2x} - e^x e^{2x} = e^x e^{2x} = e^{3x} \neq 0 \text{ for all real x.}$$

Hence, e^x, e^{2x} are linearly independent solutions of the given equation.

Clearly, $y(x) = ae^x + be^{2x}$ is also a solution of the given equation.

Now, $y'(x) = a\,e^x + 2b\,e^{2x}$.

We have $y(0) = 0$

$\Rightarrow \quad a + b = 0$

and $\quad y'(0) = 1$

$\Rightarrow \quad a + 2b = 1.$

These equations give $b = 1$ and $a = -1$.

Hence $y(x) = -e^x + e^{2x}$ is the required solution with the property that $y(0) = 0$ and $y'(0) = 1$.

Example 21:

Show that the sin x and cos x are linearly independent solutions of $y'' + y = 0$.

Solution:

If $y = \sin x$, then $y' = \cos x$ and $y'' = -\sin x$.

Thus, $y'' + y = 0$ and so $\sin x$ is a solution of $y'' + y = 0$.

Similarly, $\cos x$ is a solution of $y'' + y = 0$.

The Wronskian of $y_1(x) = \sin x$ and $y_2(x) = \cos x$ is

$$w(x) = \begin{vmatrix} y_1(x) & y_2(x) \\ y_1'(x) & y_2'(x) \end{vmatrix} = \begin{vmatrix} \sin x & \cos x \\ \cos x & -\sin x \end{vmatrix} = -1 \neq 0.$$

Hence, $y_1(x) = \sin x$ and $y_2(x) = \cos x$ are linearly independent solutions of $y'' + y = 0$.

Example 22:

Show that linearly independent solutions of

$$y'' - 2y' + 2y = 0 \text{ are } e^x \sin x \text{ and } e^x \cos x.$$

What is the general solution? Find the solution y(x) with the property; $y(0) = 2$, $y'(0) = -3$.

Solution:

The given equation can be written as

$$(D^2 - 2D + 2)y = 0 \qquad (D = d/dx)$$

Solving $D^2 - 2D + 2 = 0$, we get

$$D = \frac{1}{2}(2 \pm \sqrt{-4}) = 1 \pm i, \; i = \sqrt{-1}.$$

The general solution of the given equation is

$$y(x) = e^x (A \sin x + B \cos x).$$

We have $y'(x) = e^x (A \sin x + B \cos x) + e^x (A \cos x - B \sin x)$.

Now $y(0) = 2 \Rightarrow B = 2$.

and $y'(0) = -3 \Rightarrow B + A = -3$

$\Rightarrow A = -5$.

Hence $y(x) = e^x (2 \cos x - 5 \sin x)$ is the required solution with the property $y(0) = 2$. $y'(0) = -3$.

Clearly $y_1 = e^x \sin x$ and $y_2 = e^x \cos x$ are also solutions of the given equation. Their Wronskian is

$$w(x) = \begin{vmatrix} y_1 & y_2 \\ y_1' & y_2' \end{vmatrix} = \begin{vmatrix} e^x \sin x & e^x \cos x \\ e^x (\sin x + \cos x) & e^x (\cos x - \sin x) \end{vmatrix}$$

$$= e^{2x} \{ (\sin x \cos x - \sin^2 x) - (\sin x \cos x + \cos^2 x) \}$$

$$= -e^{2x} \neq 0 \text{ for all real } x.$$

Hence, $e^x \sin x$ and $e^x \cos x$ are linearly independent solutions of the given equation.

Example 23:

Show that $y_1(x) = \sin x$ and $y_2(x) = \sin x - \cos x$ are linearly independent solutions of $y'' + y = 0$. Determine constants c_1 and c_2 so that the solution

$$\sin x + 3 \cos x = c_1 y_1(x) + c_2 y_2(x).$$

Solution:

We have $y_2' = \cos x + \sin x$ and $y_2'' = -\sin x + \cos x$.

Thus, $y_2'' + y_2 = 0$ and so y_2 is a solution of $y'' + y = 0$.

Clearly $y_1 = \sin x$ is a solution of $y'' + y = 0$.

The Wronskian of $y_1(x)$ and $y_2(x)$ is

$$w(x) = \begin{vmatrix} y_1(x) & y_2(x) \\ y_1'(x) & y_2'(x) \end{vmatrix} = \begin{vmatrix} \sin x & \sin x - \cos x \\ \cos x & \cos x + \sin x \end{vmatrix}$$

$$= (\sin x \cos x + \sin^2 x) - (\sin x \cos x - \cos^2 x)$$

$$= 1 \neq 0.$$

Hence y_1 and y_2 are linearly independent solutions of $y'' + y = 0$. Consider the relation

$$\sin x + 3 \cos x \equiv c_1 y_1(x) + c_2 y_2(x)$$

i.e., $\sin x + 3 \cos x \equiv c_1 \sin x + c_2 (\sin x - \cos x)$.

Comparing the coefficients of sin x and cos x on both the sides, we obtain

$c_1 + c_2 = 1$ and $- c_2 = 3$.

Hence, $c_1 = 4$ and $c_2 = -3$.

Example 24:

If $y_1(x) = \sin 3x$ and $y_2(x) = \cos 3x$ are two solutions of $y'' + 9y = 0$, show that $y_1(x)$ and $y_2(x)$ are linearly independent solutions.

Solution:

The Wronskian of $y_1(x)$ and $y_2(x)$ is given by

$$w(x) = \begin{vmatrix} y_1(x) & y_2(x) \\ y_1'(x) & y_2'(x) \end{vmatrix} = \begin{vmatrix} \sin 3x & \cos 3x \\ 3\cos 3x & -3\sin 3x \end{vmatrix} = -3 \neq 0.$$

Hence, y_1 (x) and y_2 9x) are linearly independent solution of

$y'' + 9y = 0$.

Example 25(a):

Find the differential equations of:

i. the family of circles with their centers lying on the origin.

ii. the family of curves given by the equation, $y = e^{m\tan^{-1}x}$, where m is a parameter.

Solution:

i. Equation of all concentric circles with their centers lying on the origin is of the form:

$$x^2 + y^2 = a^2 \qquad \text{...(1)}$$

a being the parameter.

Differentiating (1) with respect to *x*, we get

$$2x + 2y\frac{dy}{dx} = 0 \qquad \text{...(2)}$$

∴ The required differential equation is:

$$2x + 2y\frac{dy}{dx} = 0$$

ii. $y = e^{m\tan^{-1}x}$

Differentiating with respect to *x*

$$y_1 = e^{m\tan^{-1}x}.m.\frac{1}{1+x^2}$$

$$\Rightarrow \quad (1+x^2)y_1 = me^{m\tan^{-1}x} = my$$

$$\Rightarrow \quad \frac{(1+x^2)y_1}{y} = m$$

Differentiating again with respect to x,

$$\frac{y\left[(1+x^2)y_2 + y_1.2x\right] - (1+x^2)y_1.y_1}{y^2} = 0$$

i.e., $$(1+x^2)yy_2 + 2xyy_1 - (1+x^2)y_1^{\ 2} = 0$$

Example 25(b):

Find the differential equation of the family of curves $y = Ae^{3x} + Be^{5x}$; *where A and B are constants.*

Solution:

We have, $y = Ae^{3x} + Be^{5x}$ ...(1)

Differentiating (1) with respect to x, we get

$$\frac{dy}{dx} = 3Ae^{3x} + 5Be^{5x} \qquad \text{...(2)}$$

Differentiating (2) with respect to x, we get

$$\frac{d^2y}{dx^2} = 9Ae^{3x} + 25Be^{5x} \qquad \text{...(3)}$$

$$(3) - 25\times(1) \Rightarrow \quad \frac{d^2y}{dx^2} - 25y = -16Ae^{3x} \qquad \text{...(4)}$$

$$(2) - 5\times(1) \Rightarrow \quad \frac{dy}{dx} - 5y = -2Ae^{3x} \qquad \text{...(5)}$$

$$\frac{(4)}{(5)} \Rightarrow \quad \frac{\dfrac{d^2y}{dx^2} - 25y}{\dfrac{dy}{dx} - 5y} = 8$$

$$\Rightarrow \quad \frac{d^2y}{dx^2} - 25y = 8\frac{dy}{dx} - 40y$$

$$\Rightarrow \quad \frac{d^2y}{dx^2} - 8\frac{dy}{dx} + 15y = 0,$$

which is the required differential equation.

Example 26:

Find the differential equation of the family of curves $y=(Ax+B)e^{-4x}$; *where A and B are constants.*

Solution:

We have, $y = (Ax+B)e^{-4x}$

$$\Rightarrow \quad e^{4x}y = (Ax+B) \quad \text{...(1)}$$

Differentiating (1) with respect to x, we get

$$e^{4x}\frac{dy}{dx}+4e^{4x}y = A \quad \text{...(2)}$$

Differentiating (2) with respect to x, we get

$$e^{4x}\frac{d^2y}{dx^2}+4e^{4x}\frac{dy}{dx}+4e^{4x}\frac{dy}{dx}+16e^{4x}y=0 \quad \text{...(3)}$$

$$\Rightarrow \quad e^{4x}\left[\frac{d^2y}{dx^2}+8\frac{dy}{dx}+16y\right]=0$$

$$\Rightarrow \quad \frac{d^2y}{dx^2}+8\frac{dy}{dx}+16y=0,$$

which is the required differential equation.

Example 27:

Find the differential equation of which $y=e^x\left(A\cos 2x+B\sin 2x\right)$ *is a solution; where A`and B are arbitrary constants.*

Solution:

We have, $y = e^x\left(A\cos 2x+B\sin 2x\right)$

$$\Rightarrow \quad e^{-x}y = \left(A\cos 2x+B\sin 2x\right) \quad \text{...(1)}$$

Differentiating (1) with respect to x, we get

$$e^{-x}\frac{dy}{dx}-e^{-x}y = -2A\sin 2x+2B\cos 2x \quad \text{...(2)}$$

Differentiating (2) with respect to x, we get

$$e^{-x}\frac{d^2y}{dx^2}-e^{-x}\frac{dy}{dx}-e^{-x}\frac{dy}{dx}+e^{-x}y = -4A\cos 2x-4B\sin 2x \quad \text{...(3)}$$

$$\Rightarrow \quad e^{-x}\left[\frac{d^2y}{dx^2}-2\frac{dy}{dx}+y\right] = -4(A\cos 2x + B\sin 2x)$$

$$\Rightarrow \quad \frac{d^2y}{dx^2}-2\frac{dy}{dx}+y = -4e^x(A\cos 2x + B\sin 2x)$$

$$\frac{d^2y}{dx^2}-2\frac{dy}{dx}+y = -4y,$$

which is the required differential equation.

Example 28:

Solve the following differential equations:

i. $(1 + x)\, y\, dx + (1 - y)\, x\, dy = 0$

ii. $\frac{dy}{dx} = \frac{1+y}{1+x}$

iii. $\frac{dy}{dx} + e^x + ye^x = 0$

iv. $(1+x^2)dy + x\sqrt{1-y^2}\, dx = 0$

v. $ye^{2x}dx - (1+e^{2x})dy = 0$

vi. $\sec^2 x \tan y\, dy + \sec^2 y \tan x\, dy = 0$

vii. $\frac{dy}{dx} = e^{x+y}$

viii. $\frac{dy}{dx} - xy = 0$

Solution:

i. Given that $(1 + x)\, y\, dx + (1 - y)\, x\, dy = 0$

To make the equation in the variable separable form, divide it throughout by xy.

i.e., $$\frac{1+x}{x}dx + \frac{1-y}{y}dy = 0 \qquad \ldots(1)$$

Now, (1) is in the variable separable form.

Integrating (1) we get,

$$\int\frac{1+x}{x}dx + \int\frac{1-y}{y}dy = c$$; where c is an arbitrary constant.

i.e., $\int\left(\frac{1}{x}+1\right)dx+\int\left(\frac{1}{y}-1\right)dy=c$

i.e., $\log x + x + \log y - y = c$

i.e., $(x - y) + \log (x\,y) = c$ is the required solution.

ii. Given that $\frac{dy}{dx}=\frac{1+y}{1+x}$

i.e., $\frac{dy}{1+y}=\frac{dx}{1+x}$ ...(1)

Now, (1) is in the variable separable form.

Integrating (1) we get, $\int\frac{dy}{1+y}=\int\frac{dx}{1+x}$

i.e., $\log (1+y) = \log (1+x) + c$; where c is an arbitrary constant.

i.e., $\log (1+y) = \log (1+x) + \log a$; where $\log a = c$.

i.e., $\log (1+y) = \log a(1+x)$.

or, $1 + y = a(1+x)$, or $y = a. +ax - 1$

$\therefore$ $y = ax + b$, where $b = 1 - a$ is the required solution.

iii. Given that $\frac{dy}{dx}+e^{x}+ye^{x}=0$

i.e., $\frac{dy}{dx} = -e^{x}-ye^{x}$

$\frac{dy}{dx} = -e^{x}(1+y)$

$\frac{dy}{(1+y)} = -e^{x}dx$ ...(1)

Now, (1) is in the variable separable form.

Integrating (1) we get, $\int\frac{dy}{(1+y)}=-\int e^{x}dx$

i.e., $\log(1+y)=-e^{x}+c$

iv. Given that $(1+x^{2})dy+x\sqrt{1-y^{2}}\,dx = 0$

i.e., $\frac{dy}{\sqrt{1-y^{2}}}+\frac{xdx}{1+x^{2}}=0$ ...(1)

Now, (1) is in the variable separable form.

Integrating (1) we get, $\int \frac{dy}{\sqrt{1-y^2}} + \int \frac{xdx}{1+x^2} = c$

i.e., $\sin^{-1} y + \frac{1}{2}\log(1+x^2) = c$

v. Given that $ye^{2x}dx - (1+e^{2x})dy = 0$

i.e., $\frac{e^{2x}dx}{1+e^{2x}} - \frac{dy}{y} = 0$ ...(1)

Now, (1) is in the variable separable form.

Integrating (1) we get, $\int \frac{e^{2x}dx}{1+e^{2x}} - \int \frac{dy}{y} = c$

i.e. $\frac{1}{2}\log(1+e^{2x}) - \log y = \frac{1}{2}\log a$; where $\frac{1}{2}\log a = c$

i.e., $\log(1+e^{2x}) - 2\log y = \log a$

i.e., $\log\left(\frac{1+e^{2x}}{y^2}\right) = \log a$

i.e., $ay^2 - 1 = e^{2x}$

vi. Given that $\sec^2 x \tan y \, dy + \sec^2 y \tan x \, dy = 0$

i.e., $\frac{\sec^2 x \, dx}{\tan x} + \frac{\sec^2 y \, dy}{\tan y} = 0$ ...(1)

Now, (1) is in the variable separable form.

Integrating (1) we get, $\int \frac{\sec^2 x \, dx}{\tan x} + \int \frac{\sec^2 y \, dy}{\tan y} = c$

i.e., $\log(\tan x) + \log(\tan y) = \log a$; where $\log a = c$

i.e., $\log(\tan x \tan y) = \log a$

i.e., $\tan x \tan y = a$

vii. Given that $\frac{dy}{dx} = e^{x+y}$

i.e., $\frac{dy}{dx} = e^x e^y$

$e^x dx = e^{-y} dy$ ...(1)

Now, (1) is in the variable separable form.

Integrating (1) we get, $\int e^x dx = \int e^{-y} dy$

i.e., $e^x = -e^{-y} + c$

i.e., $e^x + e^{-y} = c$

viii. Given that $\frac{dy}{dx} - xy = 0$

i.e., $\frac{dy}{dx} = xy$

then, $dy = xy\ dx$

or, $\frac{dy}{y} = x\ dx$ ...(1)

Now, (1) is in the variable separable form.

Integrating (1) we get, $\int \frac{dy}{y} = \int x dx$

i.e., $\log y = \frac{x^2}{2} + c$; where c is an arbitrary constant.

$$y = e^{\frac{x^2}{2}+c} = e^{\frac{x^2}{2}}.e^c = a.e^{\frac{x^2}{2}}, \text{ where } e^c = a.$$

[The solution is a family of exponential curves, for different values of *a*].

Example 29:

Solve the following differential equations:

i. $(x+y)dx + (x-y)dy = 0$ *ii.* $x(x-y)\frac{dy}{dx} + y^2 = 0$

iii. $(x^2 - y^2)dx + 2xydy = 0$ *iv.* $(x^2 - y^2)\frac{dy}{dx} = xy = 0$

v. $xy^2 \frac{dy}{dx} = x^3 + y^3$

Solution:

i. Given that $(x+y)dx + (x-y)dy = 0$

i.e., $$\frac{dy}{dx} = \frac{x+y}{y-x} \qquad \ldots(1)$$

It is homogeneous of degree 1.

Put $y = vx$ in (1).

This gives, $\frac{dy}{dx} = v + x\frac{dv}{dx}$

Substituting these values in (1), we get,

$$v + x\frac{dv}{dx} = \frac{1+v}{v-1}$$

i.e., $$\frac{dx}{x} = \frac{(v-1)dv}{1+2v-v^2} \qquad \ldots(2)$$

Equation (2) is in variable separable form.

$\therefore$ The solution is: $\int \frac{dx}{x} = \int \frac{(v-1)dv}{1+2v-v^2}$ ...(3)

i.e., $\log x = -\frac{1}{2}\log(1+2v-v^2) + c$ [using substitution method]

i.e., $2\log x + \log(1+2v-v^2) = \log a$; where $2c = \log a$

i.e., $x^2(1+2v-v^2) = a$ ...(4)

Now put $v = \frac{y}{x}$.

Then (4) becomes, $x^2 + 2xy - y^2 = c$, which is the required solution.

ii. Given that $x(x-y)\frac{dy}{dx} + y^2 = 0$

i.e., $$\frac{dy}{dx} = \frac{y^2}{x(y-x)} \qquad \ldots(1)$$

It is homogeneous of degree 2.

Put $y = vx$ in (1), we get,

$$v + x\frac{dv}{dx} = \frac{v^2}{v-1}$$

i.e., $$\frac{dx}{x} = \frac{(v-1)dv}{v} \qquad \ldots(2)$$

Equation (2) is in variable separable form.

$\therefore$ The solution is: $\int \frac{dx}{x} = \int \frac{(v-1)dv}{v}$ ...(3)

i.e., $\int \frac{dx}{x} = \int \left[1 - \frac{1}{v}\right] dv$

i.e., $\log x = v - \log v + c$

i.e., $\log x + \log v = v + \log a$; where $c = \log a$

i.e., $\log(vx) = v + \log a$

i.e., $vx = ae^v$ ...(4)

Substituting $v = \frac{y}{x}$, (4) becomes, $y = ae^{y/x}$, which is the required solution.

iii. Given that $(x^2 - y^2)dx + 2xydy = 0$

i.e., $\frac{dy}{dx} = \frac{y^2 - x^2}{2xy}$ ...(1)

It is homogeneous of degree 2.
Put $y = vx$ in (1), we get,

$$v + x\frac{dv}{dx} = \frac{v^2 - 1}{2v}$$

i.e., $\frac{dx}{x} = \frac{-2vdv}{1+v^2}$ ...(2)

Equation (2) is in variable separable form.

$\therefore$ The solution is: $\int \frac{dx}{x} = -2\int \frac{vdv}{1+v^2}$ ...(3)

i.e., $\log x = -\log(1+v^2) + c$

i.e., $\log x + \log(1+v^2) = \log a$; where $c = \log a$

i.e., $\log\left[x(1+v^2)\right] = \log a$

i.e., $x(1+v^2) = a$ ...(4)

Substituting $v = \frac{y}{x}$, (4) becomes,

$x^2 + y^2 = ax$, which is the required solution.

iv. Given that $(x^2 - y^2)\dfrac{dy}{dx} = xy$

i.e., $$\frac{dy}{dx} = \frac{xy}{x^2 - y^2} \qquad \text{...(1)}$$

It is homogeneous of degree 2.

Put $y = vx$ in (1), we get,

$$v + x\frac{dv}{dx} = \frac{v}{1 - v^2}$$

i.e., $$\frac{dx}{x} = \frac{1 - v^2}{v^3}\,dv \qquad \text{...(2)}$$

Equation (2) is in variable separable form.

$\therefore$ The solution is: $\displaystyle\int\frac{dx}{x} = \int\frac{1 - v^2}{v^3}\,dv$

i.e., $$\int\frac{dx}{x} = \int\left(\frac{1}{v^3} - \frac{1}{v}\right)dv \qquad \text{...(3)}$$

i.e., $$\log x = \frac{-1}{2v^2 - \log v} + c$$

i.e., $$\log x + \log v + \frac{1}{2v^2} = \log a\text{; where } c = \log a$$

i.e., $$\frac{1}{2v^2} = \log\frac{a}{vx} \qquad \text{...(4)}$$

Substituting $v = \dfrac{y}{x}$, (4) becomes, $\dfrac{a}{y} = e^{\frac{x^2}{2y^2}}$

Or, $y\,e^{\frac{x^2}{2y^2}} = a$ which is the required solution.

v. Given that $xy^2\dfrac{dy}{dx} = x^3 + y^3$

i.e., $$\frac{dy}{dx} = \frac{x^3 + y^3}{xy^2} \qquad \text{...(1)}$$

It is homogeneous of degree 3.

Put $y = vx$ in (1), we get,

$$v + x\frac{dv}{dx} = \frac{1 + v^3}{v^2}$$

i.e., $\frac{dx}{x} = v^2 dv$ ...(2)

Equation (2) is in variable separable form.

$\therefore$ The solution is: $\int \frac{dx}{x} = \int v^2 dv$ ...(3)

i.e., $\log x = \frac{v^3}{3} + c$

i.e., $3\log x = v^3 + \log a$; where $3c = \log a$

i.e., $x^3 = ae^{v^3}$ ...(4)

Substituting $v = \frac{y}{x}$, (4) becomes,

$x^3 = ae^{y^3/x^3}$, which is the required solution.

Example 30:

Use Wronskian to show that the functions, x, x^2, x^3 are independent. Determine the differential equation with these as independent solutions.

Solution:

Let $y_1 = x$, $y_2 = x^2$ and $y_3 = x^3$.

Wronskian of these function is

$$W(x) = \begin{vmatrix} y_1 & y_2 & y_3 \\ y'_1 & y'_2 & y'_3 \\ y''_1 & y''_2 & y''_3 \end{vmatrix} = \begin{vmatrix} x & x^2 & x^3 \\ 1 & 2x & 3x^2 \\ 0 & 2 & 6x \end{vmatrix}$$

$= x\,(12x^2 - 6x^2) - 1(6x^3 - 2x^3) = 2x^3 \neq 0$

$\therefore$ The given functions are independent.

To determine the equation.

Let $y = c_1y_2 + c_2y_2 + c_3y_3 = c_1x + c_2x^2 + c_3x^3$ ...(1)

$\frac{dy}{dx} = c_1 + 2c_2x + 3c_3x^2$ ...(2)

$\frac{d^2y}{dx^2} = 2c_2 + 6c_3x$ ...(3)

and $\frac{d^3y}{dx^3} = 6c_3$ ...(4)

From (4), $c_3 = \frac{1}{6}\frac{d^3y}{dx^3}$.

From (3), $2c_2 = \dfrac{d^2y}{dx^2} - 6c_3x = \dfrac{d^2y}{dx^2} - x\dfrac{d^3y}{dx^3}$ $\therefore c_2 = \dfrac{1}{2}\dfrac{d^2y}{dx^2} - \dfrac{1}{2}x\dfrac{d^3y}{dx^3}$

Subtracting x times of (2), from (1), we get

$$y - x\frac{dy}{dx} = -c_2x^2 - 2c_3x^3$$

$$= -\left(\frac{1}{2}\frac{d^2y}{dx^2} - \frac{1}{2}x\frac{d^3y}{dx^3}\right)x^2 - 2\left(\frac{1}{6}\frac{d^3y}{dx^3}\right)x^3$$

$$x^3\frac{d^3y}{dx^3} - 3x^3\frac{d^2y}{dx^2} + 6x\frac{dy}{dx} - 6y = 0$$

Which is the required differential equations.

Example 31:

Show that the Wronskian of the functions x^2, x^2 long x is non-zero. Can these functions be independent solutions of an ordinary differential equation; if so determine this equation.

Solution:

Let $y_1 = x^2$ and $y_2 = x^2 \log x$.

Wronskian of these functions is

$$W(x)\begin{vmatrix} y_1 & y_2 \\ y'_1 & y'_2 \end{vmatrix} = \begin{vmatrix} x^2 & x^2 \text{lox}\, x \\ 2x & 2x\log x + x \end{vmatrix} = x^3 \neq 0$$

$\therefore$ These functions y_1 and y_2 are linearly independent solutions of an ordinary differential equation.

To determine the equation

Let $y = c_1y_1 + c_2y_2 = c_1x^2 + c_2x^2 \log x$

so that $dy/dx = 2c_1 x + 2c_2 x \log x + c_2 x$

and $d^2y/dx^2 = 2c_1 + 2c_2 \log x + 3c_2$

Eliminating c_1, c_2 from the above three relations, we have

$$x^2 d^2y/dx^2 - 3x (dy/dx) + 4y = 0$$

which is the required differential equation.

Example 32:

Show that $y = 1/4 \sin 4x$ is a unique solution of the initial value problem $y'' + 16y = 0$ with $y(0) = 0$, $y'(0) = 1$.

Solution:

We have y' = cos 4x, y" = – 4 sin 4x.

Now, $y'' + 16y = -4 \sin 4x + 16\left(\frac{1}{4}\sin 4x\right) = 0.$

Thus, $y = \frac{1}{4}$ sin 4x is a solution of y" + 16y = 0, where

y (0) = 0, y' (0) = 1.

Since a_0 (x) = 1 ≠ 0 for each x in (–∞, ∞), it follows from Fundamental Existence Theorem that, y = 1/4 sin 4x is a unique solution of y" + 16y = 0 with y (0) = 0, y'(0) = 1.

Example 33:

Show that the functions $y = cx^2 + x + 3$ is a solution, though not unique, of the initial value problem

$$x^2y'' - 2xy' + 2y = 6$$

with $y(0) = 3,\ y'(0) = 1$ *on* $(-\infty, \infty)$.

Solution:

We have $y = cx^2 + x + 3$, $y' = 2cx + 1$. $y'' = 2c$.

$$\therefore \quad x^2y'' - 2xy' + 2y = x^2(2c) - 2x(2cx + 1) + 2(cx^2 + x + 3)$$

$$= 4cx^2 - 4cx^2 - 2x + 2x + 6 = 6.$$

Thus, $y = cx^2 + x + 3$ is a solution of $x^2y'' - 2xy' + 2y = 6$,

where y (0) = c × 0 + 0 + 3 = 3, y' (0) = 2c × 0 + 1 = 1.

Here $a_0(x) = x^2$, $a_1(x) = -2x$, $a_2(x) = 2$, $R(x) = 6$ are continuous functions on (– ∞, ∞)

It may be noted that a_0 (x) = 0 for x = 0 ∈ (– ∞, ∞).

Hence the solution $y = cx^2 + x + 3$ is not unique, as $y = cx^2 + x + 3$ is a solution for any real value of c.

Notice that $y = x^2 + x + 3$, $y = 2x^2 + x + 3$ are both solutions of $x^2y'' - 2xy' + 2y = 6$ with y(0) = 3, y'(0) = 1.

Example 34:

Find the differential equation of all tangent lines to the parabola

$$y = x^2.$$

Solution:

Parametric equations of the given parabola are

$x = t, \qquad y = t^2$

$\therefore$ The equation of any tangent is

$$2xt = y + t^2 \qquad \text{...(1)}$$

Equation (1) is the family of tangent lines to the parabola, where t is the parameter.

Differentiating (1) once

$$2t = y^{(1)} \qquad \text{...(2)}$$

Eliminating t between (1) and (2)

$$4(y - xy^{(1)}) + (y^{(1)})^2 = 0$$

Form the differential equations for the following families of curves.

Example 35:

Show that $y_1(x) = \sin x$ and $y_2(x) = \sin x - \cos x$ are linearly independent solutions of $y'' + y = 0$.

Solution:

Both $y_1(x) = \sin x$ and $y_2(x) = \sin x - \cos x$ satisfy the differential equation $y'' + y = 0$.

$\therefore$ They are solutions of the equation.

Now Wronskian of these solutions y_1 and y_2 is

$$W(x) = \begin{vmatrix} y_1 & y_2 \\ y'_1 & y'_2 \end{vmatrix} = \begin{vmatrix} \sin x & \sin x - \cos x \\ \cos x & \cos x + \sin x \end{vmatrix}$$

$$= \sin x (\cos x + \sin x) - \cos x(\sin x - \cos x) = 1 \neq 0.$$

$\therefore$ These $y_1(x)$ and $y_2(x)$ are linearly independent solutions of $y'' + y = 0$.

EXERCISES

1. Show that the set of functions

   e^{ax}, e^{bx}, e^{cx} (a, b, c distinct) is linearly independent for all x.

2. Show that the following functions are linearly independent:

   $e^x \cos x, e^x \sin x$

   From the differential equation of second order having these two functions as independent solutions.

3. Show that the Wronskian of the function

   $x^a, x^b, x^c, (x > 0)$ is equal to $(a - b)(b - c)(c - a)\, x^{a+b+c-3}$

4. Which of the following sets of functions are linearly independent for all x?

(i) sin x, cos x,

(ii) sin 3x, sin x, $\sin^3$ x,

(iii) sin bx, e^x, e^{-x}

(iv) e^x, xe^x, sinh x.

5. Solve the following differential equations:

   i. $(2x+3y)dx+(3x-2y)dy=0$

   ii. $2xydx+(x^2+4y)dy=0$

   iii. $(ax+hy+g)dx+(hx+by+f)dy=0$

   iv. $(e^y+1)\cos x\ dx+e^y\sin x\ dy=0$

   v. $x(x^2+y^2-ax)dx+y(x^2-y^2+by)dy=0$

6. Solve the following differential equations:

   i. $\frac{dy}{dx}+y\cot x=e^x$ ii. $\frac{dy}{dx}+y=x$

   iii. $\frac{dy}{dx}+\frac{y}{x}=\frac{1}{x}$

7. Solve the following differential equations:

   i. $\frac{d^2y}{dx^2}+5\frac{dy}{dx}+6y=0$ ii. $(4D^2-6D+8)y=0$

   iii. $\frac{d^2y}{dx^2}+6\frac{dy}{dx}+9y=0$ iv. $\frac{d^2y}{dx^2}+y=0$

   v. $(D^2-4D+5)y=0$

8. Solve the following differential equations:

   i. $\frac{d^2y}{dx^2}-7\frac{dy}{dx}+6y=e^{2x}$ ii. $(D^2+4D+4)y=e^{-3x}$

   iii. $\frac{d^2y}{dx^2}+6\frac{dy}{dx}+9y=50e^{2x}$ iv. $(2D^2+4D+3)y=2e^x+e^{-x}$

   v. $(D^2+a^2)y=e^{ax}+e^{bx}$ vi. $\frac{d^2y}{dx^2}-9y=1+5e^{4x}+2e^{3x}$

   vii. $(4D^2+16D+15)y=4e^{-3x/2}$

   viii. $\frac{d^2y}{dx^2}-6\frac{dy}{dx}+8y=(e^{2x}-1)^2$

9. Solve the following differential equations:

i. $\frac{d^2y}{dx^2}+9y=e^x-\cos 2x$

ii. $(D^2+1)y=e^x+e^{-x}+\sin 2x$

iii. $\frac{d^2y}{dx^2}+4\frac{dy}{dx}+4y=e^{3x}+\cos 5x$

iv. $(D^2-2D-8)y=4\cos 2x+e^{4x}$

v. $(D^2-3D+2)y=\sin 3x$

vi. $\frac{d^2y}{dx^2}-4\frac{dy}{dx}-5y=4\cos 3x+e^{3x}$

vii. $(D^2+3D+2)y=2e^{-x}+\sin 2x$

viii. $\frac{d^2y}{dx^2}+4y=4\cos 2x$

10. Solve the following differential equations:

i. $(D^2+D+1)y=x^2$

ii. $(D^2+3D+1)y=3x^2$

iii. $\frac{d^2y}{dx^2}-y=3+2x^3$

iv. $(D^2+3D)y=6x+5$; given that $y=0, \frac{dy}{dx}=0$, when $x=0$

v. $(D^2-3D+2)y=x^2+e^{4x}$

vi. $\frac{d^2y}{dx^2}+9y=3x+e^{2x}$

vii. $(D^2+3D+7)y=x+e^{2x}$

viii. $\frac{d^2y}{dx^2}+\frac{dy}{dx}-2y=x+\sin x$

ix. $\frac{d^2y}{dx^2}+4y=\sin 2x+x^2$

x. $\frac{d^2y}{dx^2}+16y=\cos 4x+x^2$

11. Form the differential equation by eliminating A and B from:

    $y = Ae^{3x} + Be^{-x}$

12. Evaluate the Wronskian of the functions e^x, xe^x. Hence conclude whether or not they are linearly independent. If they are independent, set up the differential equation having them as its independent solutions.

13. Show that any two solutions of the equation

    $y'' + f(x)\, y' + g(x)\, y = 0$

    f (x) and g (x) are continuous on an open interval 1, are not linearly independent, if and only if, their Wronskian is zero for some $x = x_0$ on 1.

14. Define Wronskian. Prove that the functions 1, x, x^2 are linearly independent. Hence from the differential equation whose roots are 1, x, x^2.